Timber

Timber

Peter Dauvergne and
Jane Lister

polity

First published in 2011 by Polity Press
Reprinted in 2011

Polity Press
65 Bridge Street
Cambridge CB2 1UR, UK

Polity Press
350 Main Street
Malden, MA 02148, USA

ISBN-13: 978-0-7456-4927-6
ISBN-13: 978-0-7456-4928-3(pb)

A catalogue record for this book is available from the British Library.

Typeset in 10.25 on 13 pt FF Scala
by Servis Filmsetting Ltd, Stockport, Cheshire
Printed and bound in the United States by Odyssey Press, Inc., Gonic, New Hampshire

For further information on Polity, visit our website: www.politybooks.com

Contents

Figures, Tables, and Boxes

TABLES

BOXES

Acknowledgments

Many people provided helpful consultation during the drafting of this book, including Graeme Auld, Gary Bull, Linda Castagna, Lars Gulbrandsen, Robert Kozak, Anna Tikina, Ben Toosi, Antje Wahl, Mark White, and Peter Wood. We are grateful to the three anonymous academic reviewers. As well, we would like to acknowledge several additional readers for their generous and insightful reviews of the manuscript, including Bill Cafferata, Linda Coady, Peter Gill, Bruce Lister, Kate Neville, and Natalia Vidal. The book would not have been possible without the patience and support of the staff at the Liu Institute for Global Issues: Rita Zamluk, Patty Gallivan, Sally Reay, Tim Shew, and Julie Wagemakers. And, finally, we would like to thank Louise Knight and David Winters at Polity Press for the opportunity to contribute to this innovative series on the global geopolitics of resources and for their supportive professional guidance throughout the publication process.

CHAPTER ONE

The Global Political Economy of Timber

It is hard to get through a day without relying on timber. It is one of the world's most versatile natural resources used in everything from home construction, furniture, packaging, books, diapers, hospital gowns, and currency to paint solvents, food, pharmaceuticals, and fuel. It is also one of the world's most renewable resources. Unlike oil, we will not run out of timber. Brazil alone could meet the world's total timber demand. Unlike other natural resource crises, adequate supply is not, nor will it become, the critical issue. Rather, the global challenge of timber is about the increasing loss of ecological services, forest biodiversity, and community well-being from the way timber is now logged, traded, produced, and consumed through globalizing commodity chains – production and consumption pathways that are lengthening and multiplying across and between continents as multinational discount retailers like Walmart, Home Depot, and IKEA increasingly turn to the developing world as a source of cheap products to maintain profits and serve bargain-hunting consumers.

Ten thousand years ago, vast, contiguous natural forests blanketed much of the world. The history of timber as a resource worth fighting for began as settled agricultural communities cleared these forests, as towns built forts, and as schooners battled at sea. Yet, even as monarchs took control of timber in some places, in most of the world it remained a shared community resource. Here, few had any reason to fight over what was a renewable and seemingly inexhaustible asset.

Today, more than half of these original forestlands are gone, with deforestation since 1950 roughly equal to all of the previous loss and with many natural forests fragmented or replaced by industrial timber plantations. For temperate and most boreal forests, governments from China to the United States are now enforcing stricter logging rules, reforesting degraded land, and extending parklands. Many problems remain. But, on the whole, the last few decades have seen the management of forests in strong regulatory states – those with a capacity to enforce rules consistently – gradually improve on at least some measures, such as reforestation rates.[1]

At the same time, management of many tropical forests as well as the Russian boreal forest remains as bad, or is worsening, and deforestation is a greater threat than ever before. Here, some of the poorest communities and most vulnerable ecosystems – in places like the Amazon, Borneo, Siberia, and sub-Saharan Africa – are in a spiraling decline.

From a purely ecological stance some might wonder if the gains in the temperate world might offset the declines in tropical and Russian boreal forests. The answer is no: not even close.

Tropical regions are experiencing sweeping forest fires, extensive soil erosion, and flash-flooding. Slow-growing boreal regions are experiencing vast clear-cutting. The net result is an accelerating decline in the earth's capacity to support communities, sustain biodiversity, and store carbon dioxide. Undisturbed tropical forests alone absorb more than one billion tonnes of carbon from the atmosphere every year. Deforestation, meanwhile, is one of the biggest contributors to climate change, accounting for as much as one fifth of anthropogenic carbon dioxide emissions – more than the global transport system.[2] This adds as well to inequality and insecurity as deforestation is greatest in some of the world's lowest-income communities and most unstable states, where forest loss contributes to rural violence and the spread of disease, among many other problems.

Why is this happening? The broad, encompassing answer is poverty and population pressure. Forests are increasing in high-income countries with stabilizing populations, whereas by necessity developing countries with rising populations are seeking income gains by harvesting their forestland and converting it to higher-value economic uses. Development, expanding cities, growing markets, and rising per capita consumption in emerging economies are all putting greater demands on forest resources. A closer analysis, however, reveals that the causes of tropical deforestation are not strictly within the developing world, but lie as well in high-consuming developed countries.

Fundamentally, an underlying reason for continued deforestation comes down to who is controlling and consuming timber, and how and why this is changing. Our book zeros in on the consequences of one factor in particular: the globalization of corporate commodity chains that move products from producers to consumers. Changes *within* these chains, we argue, are increasing the pressures on the remaining undisturbed native forests and degraded forestland of the Third World (what we also refer to as the "global South," or sometimes simply "the South"). Significant international capital investments are now flowing into large-scale modern pulp and paper and solid wood manufacturing facilities in the South (such as the emerging economies of Brazil and China). To some extent the new mills (many in partnership with firms headquartered in "the North") are supplying emerging domestic markets; but most are aiming to supply customers in the North, who already consume around three-quarters of the world's solid wood and close to two thirds of its paper. Moreover, a significant quantity of timber "consumed" in the South ends up on the retail shelves and landfills in the North, not shipped as "products" but as paper packaging and pallets to wrap, protect, and brand growing exports of manufactured goods.

The rapid rise in global timber production capacity in the South has created large regional fiber deficits (especially in Asia). This has led to a frenzied demand for cheap – and often illegal – tropical and Russian boreal timber to feed the new mills. To help meet the growing demand, governments and companies have been establishing fast-growing industrial timber plantations. Often, though, these are at the expense of soil and water quality, native forests, and forest-dependent communities (sometimes displacing people). And multi-national timber corporations now rely on shipping low-cost pulp, logs, lumber, paper, and manufactured wood products back and forth across the globe in an effort to gain cost advantages, maximize profits, and maintain market shares.

Third World suppliers, by integrating into these globalizing timber chains, can make lots of money. Yet, to meet low-price contracts, many end up working harder and harder to keep costs as low as possible – strategies that include logging destructively or illegally and evading taxes or fees. Consumers in the North reinforce the system by buying timber products – furniture, flooring, cribs, picture frames, paper – at quantities and prices that largely ignore the environmental and social costs. Moreover, these bargain products are packaged in corrugated boxes and loaded onto millions of wooden pallets (most of which end up in a landfill after a single use) – all of which further drives up wasteful, unsustainable timber consumption. The net effect is to cast deeper and longer ecological and social shadows of high consumption in the First World onto poor peoples and fragile environments in the Third World.[3]

The future of intact, old-growth tropical forests does indeed look bleak. These biologically rich ecosystems are now on a path to disappear by the end of the century.[4] And more efficient manufacturing, rising rates of paper and wood recycling, and expanding timber plantations are doing little to slow this loss. In many cases industrial plantations in the global South are making conditions even worse – doing little for the local environment as fast-growing species like eucalyptus are pulped into "luxury" exports, such as "soft" and "bright" toilet paper. Yet, intriguingly, as we explore in the book's conclusion, the shifting power balances among corporations within global commodity chains are also creating some possibilities to leverage the power dynamics within these chains to improve logging and timber manufacturing practices in developing countries.

Corporate powerhouses

Multinational forest and paper corporations from Europe, North America, and Japan are still at the core of the global timber industry. Yet, as the Food and Agriculture Organization (FAO) explains, the model of a large vertically integrated company is "giving way to highly networked global supply chains, linking firms and affiliates across countries."[5] Giant retail chains and wood product buyers like Home Depot, Lowe's, IKEA, and Walmart are expanding in size and global reach, and changing the structure and rules of global timber supply chains with their growing "buyer-driven" power to control suppliers.[6] This shift to greater retail control and more networked supply chains is also providing opportunities for Third World firms to grow rapidly and capture global market share. The consequence is the erosion of some of the historical control of First World (Northern) timber multinationals, although in terms of market capitalization and sales Third World timber companies remain significantly smaller with a relatively weak presence in higher-end First World markets.

With profit margins narrowing in mature home markets

(ones with limited new growth potential) over the last few decades, the traditional Northern forest and paper multinational corporations (MNCs) have been partnering and relocating manufacturing facilities to the Third World in search of cheap fiber (both plantation and natural) and lower operating costs. Commodity chains supplying consumers from New York to Tokyo have become longer and more complicated, increasingly flowing through every emerging national economy – including Brazil, China, India, Indonesia, and Russia – as well as through every developing region. Trade no longer flows primarily along a consistent and direct pathway from producer to consumer. Most products instead move through distributed and multidimensional networks of ever-shifting players and multiple owners along obscure routes that crisscross borders and markets.

Northern timber manufacturers and retail companies retain most of the power within these chains. They especially have control over structural features that shape the means of corporate interaction and the range of options for all actors (such as legal contracts, purchasing agreements, and performance standards). These companies also tend to frame – and thus largely control – the private and semi-private policies and discourse (language, ideas, and values) within and around these chains, such as "voluntary regulation" and policies like corporate social responsibility (CSR). This form of power is more indirect, although often no less (and sometimes more) effective, as it constitutes and can precede more formal corporate contractual arrangements.[7] Governments and international institutions such as the World Trade Organization and the World Bank can further enhance the legitimacy of this framing – and in turn the structural power of Northern companies – by endorsing and reinforcing this corporate discourse.

The rise of a discount economy of mass production and merchandizing of cheap products and the strengthening of global retail chain power are mutually reinforcing trends. Big retail can wield a big stick: few Third World loggers or timber processors can afford to be replaced by another supplier. "Switching," in the language of business, can be fairly easy and inexpensive to do for many standardized timber products, especially for a mega-retailer like Walmart.

For this reason a company like Walmart, which commonly accounts for 10–30 percent of an average supplier's business, has considerable capacity to influence the actions of its suppliers: potentially far more, for example, than a weak regulatory state such as Indonesia where politicians and military officers often shield timber firms. To some extent this is even true in a relatively strong emerging economy like China. Benny Fung, managing director of Lutex, a Hong Kong-based health and beauty company, explains this nicely in the case of Walmart: "We heard that in the future, to become a Walmart supplier, you have to be an environmentally friendly company. So we switched some of our products and the way we produced them."[8]

So far, however, few of the large discount retailers are consistently exercising their power to raise environmental or social standards. Instead, they are more likely to use it to squeeze their suppliers to leverage a better price deal. Big box retailers, for example, are known to strike buying contracts with suppliers that take up a large fraction of a factory's production capacity. The supplier then invests in a redesign of its facility to meet the huge order. The following year the retailer may then offer the same purchase order, but now demand a lower price: say, a 5 percent discount.

The supplier is then stuck, often having gone into debt to raise production to meet the first year's big order. A firm caught in this situation generally cannot afford to not comply – somehow, cost reductions must be found. The result can be innovation, but usually not what one thinks of as responsible

business practices. Rather, the creative energy of the corporate executives goes into finding ways to cut corners and hide
illegalities and poor business practices: for example, by falsifying documents and audit reports, misreporting tree species
and volumes to reduce government fees, or creating "front-
companies" to disguise the real manufacturing source of the
products sold to the big box retailer.

Companies headquartered in the North rely also on Third
World suppliers as agents for the exercise of more blunt
forms of power: from coercing locals to bribing officials to
evading taxes. Suppliers from the global South are not, however, mere pawns of the North. There is considerable variation
in the organization and operation of chains across the wide
array of timber products.[9] All Third World suppliers wield at
least some power within these chains. And a few, especially
on a regional scale, are even chain leaders. The growing
market capacity of firms like Fibria in Brazil and Shandong
Chenming Paper in China, for example, is to some extent
eroding the historical control of Northern timber MNCs over
global trade and markets.

One sign of the power of some "Southern" suppliers is
the move over the last decade by Northern companies – both
timber producers and retailers – to partner with suppliers
from emerging economies such as Brazil, Russia, Indonesia,
and, in particular, China so as to gain access to new markets,
low-cost production facilities, large workforces, and inexpensive inventory. Even without significant power over chain
structures or corporate discourse, some Third World firms are
now rising up the corporate hierarchy of profits and market
control at a much faster rate than the traditional Northern
MNCs did over the last century. These firms are doing this
by taking advantage of cheap local wood sources, low operating costs, and opportunities to integrate into buyer-driven
commodity chains in more complementary – rather than

principally competitive – relationships with integrated forest and paper companies and big retailers.

Corporate control over the world's forests arises, then, from a lively competition among firms jockeying for technologies, profits, and markets within global commodity chains. The power to do so, however, arises in part from the commoditization and distancing of forest value: where trees are primarily valued as a commercial resource to trade *globally* rather than as an ecosystem or home for local communities.

Corporatized forests

Corporations worldwide have gained power over forests: not so much by fencing natural forests off from local people, who still rely on large volumes of branches and small logs for fuel, but more by controlling the profits from processing, trading, and retailing trees as a commercial resource in a globalized market. Governments have more than acquiesced to the corporatization of forests. Although more than 80 percent of the world's forests remain publicly owned, governments have handed large concessions to big timber companies in exchange for financial rent and employment. In hindsight, these decisions frequently look shortsighted as the benefits are often temporary, in the absence of strong forest regulations only lasting until all of the commercially valuable trees are cut down.

Fuelwood still accounts by volume for about half of global wood consumption. But the global value of forests is no longer mainly as a source of wood for cooking, heating homes, or local building, as was the case for most of the world even a hundred years ago. Nor do they have much global value as places of worship, for spirits to reside, or even for indigenous people to live. Forests from Brazil to Indonesia and from Siberia to Liberia now supply the world with a global commodity called "timber."

Many of us picture here a mahogany desk or a cedar house. But timber – or, in more technical terms, industrial roundwood – goes into making a great variety of products.[10] By value, half of the world's timber goes to make paper and paperboard. Much of this produces things that do not immediately bring to mind a timber product: diapers, milk cartons, toilet paper, and poker cards. As well, half of the world's paper goes into packaging. Paper packaging protects and brands food, consumer goods, and manufactured parts – from cereal, milk, mangoes, and basketballs to computers, TVs, refrigerators, and auto parts.

Corporations have even more control over forestland in weak regulatory states. Again, this control is not primarily over natural forests as territory or property, but over the commercial value from exploiting these lands. Violence to strong-arm locals out of the way is definitely part of the toolkit of some loggers. And in places like Southeast Asia loggers are commonly protected from prosecution by politicians, military officers, and the police. But corporate land grabs with the goal of long-term control are far more common in the case of agricultural or timber plantations than in the case of natural forests.

For logging companies operating in natural forests in the South, it is often more important to gain short-term access rather than retain lasting control. Because of complex land-tenure conditions, loggers across the global South in fact often have no option but to hold overlapping, informally defined "legal" timber concessions as local patrons hand out – or revoke – licenses. Those who actually log the forests range considerably. Some are locals hired by firms or traders. Some are private, family-owned firms with a few dozen employees. Some are major Third World companies such as Fibria in Brazil. Some are company affiliates under large umbrella holding companies like Malaysia's Rimbunan Hijau and Indonesia's Sinar Mas Group. And some are MNCs with worldwide operations, such as International Paper (US), Stora Enso (Sweden/Finland), and Oji Paper (Japan).

Regardless, most commercial loggers in these low-income weak states have little interest in long-term management of "their" forestland; instead, they simply harvest as much as possible as quickly as possible. Logging tends to ramp up and become even more destructive when timber prices are high or rising. Mounting uncertainty in future access to a legal concession – from, say, political unrest or an upcoming election – tends to speed up harvesting even more, with no regard for the 30- or 40- or 50-year plans on paper to produce "sustainable timber yields." Loggers generally abandon a site when it runs out of the most profitable commercial timber – or when prices crash, a common feature of the world timber market. Here, a frequent result is an environmental pattern of cut-and-run in an economic cycle of boom-and-bust. The degraded forestland then becomes a target for conversion – mainly for industrial agricultural plantations such as soy, oil palm, or fast-growing tree species such as eucalyptus, pine, and poplar.

Regulations for "sustainable forest management" may look reasonable on the books: yet, with little to no enforcement, much of the so-called "legal" logging is highly destructive. Log prices tend to reflect the cost of cutting down a tree and getting it to a mill, and not any of the longer-term – and much bigger – environmental and social costs. Moreover, by allying with local leaders and corrupt politicians, loggers and traders are able to evade taxes and fees, harvest illegally, and smuggle timber into the global market. This increases corporate and personal profits while further lowering consumer prices and government revenues. It also leaves already poor places even poorer as employers move on without addressing the long-term environmental and social disruption.

In recent decades this process has been accelerating and deepening even further as the globalization of corporations, financing, trade, and consumption increasingly integrates resource-dependent economies into ones more central to the global economy.

Globalizing timber chains

Timber, like water and food, has always been an essential resource for people everywhere, with only the driest, highest, and coldest lands devoid of forests. Some societies, such as the Phoenicians of 1200 BC and the Minoans on Crete in 1450 BC, ran out of timber. Others, such as the British Empire from the early 1600s to the mid-1900s, enhanced their power by colonizing forests from North America to Africa to South and Southeast Asia. In some ways the globalization of timber production and consumption started during the era of the French, British, and Dutch colonists. Still, even a half-century ago most societies, particularly those in the developing world, obtained sufficient timber supplies by trading wood locally – or at most regionally – from nearby forests.

Today, the picture is totally different. Of the world's 4 billion hectares of boreal, temperate, and tropical forests half are now set aside for "production" (for both direct and multiple purposes). Timber commonly travels across many countries, as it moves through global commodity chains. At least 5,000 wood products are now trading back and forth from South to North and North to South, often many times. A company in Indonesia might turn a log into lumber or pulp and then trade it to another company in China. That company might then turn it into furniture, flooring, or paper, and then sell it to retailers in North America and Europe. Even this would be a very simple commodity chain, however: most involve far more transactions across far more jurisdictions.

For much of the twentieth century, companies with a home base in a heavily forested Northern country – ones like America's International Paper, Georgia-Pacific, and Kimberly-Clark, Sweden's Svenska Cellulosa AB (SCA), and Finland-Sweden's Stora Enso – dominated the global timber industry. Such companies are still powerhouses, providing fiber, financing, equipment, and markets. Yet their power is diffusing as globalization integrates more developing countries into the world economy, and as the source of logs shifts from the temperate and boreal forests and plantations of the First World to tropical old-growth forests, Russia's boreal forests, and fast-growing plantations in Asia, Africa, and Latin America.

Processing the global South

Few plantations existed 50 years ago and they supplied very little timber. Now they cover hundreds of thousands of hectares and provide about one third of the world's industrial wood fiber – and this amount is gradually increasing. The southern United States remains a core source of the world's plantation wood; but plantations are also expanding quickly in Russia and China. As a result, more than half of the world's timber plantations are in just three countries: China, the US, and Russia. The fastest-growing and lowest-cost timber plantations, however, are in South America (e.g., Brazil and Chile).

Over the last decade, China in particular has been investing aggressively to increase domestic capacity to produce plantation timber. And the future will likely see even more plantations in the global South as companies seek out inexpensive land and overhead. The capacity of corporations to clear vast areas of the South – such as burning down degraded tropical forests – and then control this land (often with government assistance) will further reinforce this likely trajectory.

The processing of logs – whether from plantations or

natural forests – occurs in a great variety of places: from paper factories outside Shanghai to furniture makers in the Brianza region of Italy to pulp mills in the Bahia state of Brazil to plywood mills outside Jakarta. As in the case of log production, however, over the last few decades the process of globalization has been shifting timber processing into emerging economies, such as Brazil, Chile, China, Indonesia, Vietnam, and Russia, and away from traditional First World processors, such as Canada, Sweden, Finland, the US, and Japan. This trend is partly a result of Northern companies investing in Third World timber-processing facilities and pulp and paper mills to reduce costs and access new log supplies (both natural and plantation). A significant driver of log processing in the South is also the decision of many Third World states to ban raw log exports (or impose prohibitive tariffs) to create jobs and diversify economies and encourage value-added log processing "at home."[11] Despite the log export bans, however, daily truckloads of tropical and boreal logs still rumble across the borders of many states. In many cases, these logs are being smuggled out. Overall, though, the majority of timber traded, especially the legal trade, is semi-processed or processed rather than raw logs.

Many of today's Third World timber processors – such as Klabin and Suzano in Brazil, Masisa in Chile, Ballarpur in India, and Yuen Foong Yu Paper and Cheng Loong in Taiwan – are going global by integrating into a global timber chain. From Asia to Africa, small producers are starting up or reorganizing within these chains to gain access to foreign direct investment, new technologies, and higher-end markets.

In an increasing number of cases they are ending up as part of a corporate network with a big retailer at the helm, providing them with specifications for everything from kitchen cabinets and shelving to doors and decking. Local suppliers unable, or unwilling, to integrate into a global commodity chain tend to lose market share, or go out of business. Meanwhile, some of those integrated into a chain are experiencing unprecedented growth: unlike most Northern timber companies that took a hundred or so years to become world leaders, some in the global South are now vaulting up the corporate rankings in as little as a decade or so.

The rise of China

The rise of China over the last decade as a timber importer and processor is especially notable. China is now the world's second biggest wooden furniture maker and paper producer (after the United States). Since 1990, China has accounted for around half of the global increase in paper production. Such rapid growth has demanded massive wood fiber inputs. China's timber imports quadrupled in the decade following a 1998 ban on logging in its natural forests. And about half of the world's traded timber now goes through or lands in China. Much of this is unsustainably and illegally produced with some also entering unlawfully – with the "illegal" portion ranging from 30–80 percent from countries like Russia, Indonesia, Burma, and Papua New Guinea.[12]

Partnering with Chinese firms has become vital in the struggle for a share of world timber profits. And all of the world's biggest timber MNCs have now integrated low-cost Chinese mills and factories into their supply chains. State companies – along with numerous small local trading houses – historically controlled China's timber importing and exporting. The government and local traders remain significant, but the opening of China's economy has provided private companies with more direct access to overseas timber supplies and markets. Networks of firms have formed commodity chains that link the 200,000 or so small- to medium-scale manufacturers within China's fragmented wood sector to timber suppliers in low-cost regions (e.g., Russia and Southeast Asia), as well as to buyers in large, increasingly concentrated Northern consumer markets.

China's domestic consumption of timber is booming, too. Yet, with per capita consumption still far below developed countries, much of China's timber imports are further processed, then exported, with the final purchaser more often residing in the First World. Northern MNCs have shown great interest in linking into commodity chains where high-quality logs and lumber are shipped to Chinese mills for processing, then exported back to big retail customers in the North. Such business partnerships and supply-chain relationships within China, however, are not limited to Chinese manufacturers.

Over the last decade timber manufacturers from countries like Japan, Taiwan, the US, and Europe have all relocated facilities to China – not just to compete within China's domestic market, but also to make low-cost products for export overseas. The relocation and outsourcing by American firms, for example, has been vital for China's meteoric rise as a wood furniture producer and exporter. More than one third of China's furniture exports go to the United States; and China now supplies more than half of total US furniture imports (by value). China is also the main source of furniture for companies such as IKEA and Walmart. Official export figures for consumer items like furniture, moreover, only capture a portion of China's timber "exports." Vast amounts of timber leave the country as "free" packaging and pallets to wrap, ship, and brand goods.

A tough business

China is faring remarkably well in what remains a tough business. Loggers work in remote and dangerous locations. Competition among firms is fierce. And timber prices often swing suddenly and widely. Windfall profits, not unlike during a gold rush, are certainly common; but many firms go bust as well.

The global financial downturn of 2007–09 added further to industry woes, forcing many timber companies to restructure or go bankrupt. Some Third World timber companies fared well, however, with a few recovering strongly in 2009. After a decline in 2008, for example, the stock performance of China's two biggest packaging companies (Hong Kong-based Nine Dragons Paper and Lee & Man Paper) rose 398 percent and 393 percent respectively in 2009, even as the financial performance of the Northern timber companies was relatively poor (the American-owned Pope & Talbot Company, for example, went bankrupt in 2008 after 160 years in operation).

To gain or maintain a competitive advantage within this challenging – and globalizing – business setting, companies are increasingly merging and cooperating, with deep consequences for the current and future geopolitics of timber. The big multinationals – through financing, purchasing, and facilitating logistics – wield considerable power in the resulting global commodity chains. MNC support is especially vital for small-scale loggers and processors in the global South. International institutions – most notably the World Bank, International Monetary Fund, and World Trade Organization – also reinforce these business networks through lending, business advice, and structural adjustment and trade liberalization policies that encourage Third World governments to open their economies to powerful Northern companies, as well as to provide investors with tax breaks, devalue currencies, and create environmental loopholes. The need to service foreign debts also tends to encourage developing countries to implement policies that promote timber exports.

Maintaining a toehold in a commodity chain requires smaller companies to work on the principle of "low-cost at all cost." Heavyweights like Home Depot, Lowe's, IKEA, and Walmart wield great power over the extent and distribution of profits within these discount product chains. Quality and design of products do matter for these mega-retailers. But their principal focus is on moving discounted items, in bulk, fast. Profit margins per product or transaction are intentionally low.

The goal is high net profits through high-volume sales: in other words, as quickly as possible moving a lot of inventory to a lot of consumers. Part of these profits comes from undervaluing the environmental and social costs of producing timber. And part comes from advanced logistics technology and low global shipping costs. Companies also take advantage of built-in government subsidies. These include access to public infrastructure, such as heavily subsidized roads to truck goods "just-in-time" through the commodity chain.

By aggregating world consumer demand for cheap products, these retailers enable – indeed, actively encourage – bargain-hunting by consumers. This in turn spins back up the commodity chain, further increasing the pressure on suppliers to cut corners, mistreat employees, and log destructively and illegally to keep costs down – and thus allow the big box stores to offer consumers "unbeatable" prices. Retailers in the forest sector wield particularly great buying power in timber supply chains because, as we mentioned earlier, the costs of switching suppliers – regarding lost customer loyalty, termination penalties, transaction costs of finding new suppliers, or losing regional economies of scale – tend to be low. Eagerly waiting in the wings are many forest and paper suppliers who can produce similar products. This makes it easier for these retailers to replace their suppliers with few repercussions.

More broadly, these companies also reinforce (and are part of) a discount market economy and global culture of "disposable consumerism," where people put less value on craftsmanship and things handed down over generations, and instead value the convenience of owning do-it-yourself assembly furniture that is so inexpensive (yet still fashionable and functional) that it makes financial sense to throw it away rather than move or repair it. The globalization of trade and the growth of online retail are also driving up the global consumption of packaging – more than half of which quickly ends up as landfill.[13] All of this is contributing to more wasteful and costly consumption of the world's timber.

The rising costs of global timber consumption

An increase in the consumption of plastic, concrete, and steel, although hardly a good environmental trend, has helped to counteract some of the pressures to consume more timber. Wood processing itself is also more efficient and innovative now, allowing companies to produce a wider range of engineered and finished products using less timber. Recycling of wood fiber – especially paper and paperboard – has been increasing steadily as well in recent years. Bamboo and rattan as well as plantation trees are helping to meet some of the rising demand for timber, too.

For natural forests, however, industrial plantations are a double-edged trend. In a lot of places, particularly in the global South, demand for commercial plantation land is driving deforestation. Plantation timber as a "substitute" is also not a substitute for the ecological or social value of a natural forest. In many ways "forest" plantations are not much different from any agricultural plantation: relying on regular clearing of monoculture crops grown using irrigation, fertilizers, and pesticides.

Overall, these counteracting trends have not stopped the consumption of timber from natural forests from continuing to rise. Increases in under-regulated harvesting entail big costs, adding further to the irreparable degradation of forests, loss of biodiversity, and climate change. Globalizing commodity chains are hiding, distancing, and obscuring many of these costs from end consumers. Timber prices that externalize the environmental and social costs of logging further contribute to wasteful and excessive consumption. The effect is to outsource much of the forest loss from producing timber for consumers in the global North to the poorest countries and the poorest peoples.

Many consumers are largely unaware – and corporations largely unaccountable – for these shadow effects of consumption. How many consumers in Japan, for example, would connect living in a concrete building to deforestation in Southeast Asia and Melanesia? Yet over the last half-century the most common use of the giant old-growth trees of this region has been for plywood paneling to mold concrete in Japan. Called *kon pane* in Japanese, construction companies normally burned or discarded these panels after just a few uses. Why such incredible waste? The answer is simple: it was cheaper to buy new panels than clean the old ones.

Rising consumption of products growing on deforested land is also creating powerful incentives to clear forestlands. Increasing global consumption of margarine and oil for deep-frying, for example, is encouraging companies to burn down forests on Indonesia's outer islands to clear land for oil palm plantations. Increasing demand for palm oil as a biofuel substitute is starting to drive forest conversion, too.

Forest fires in Indonesia, which now rage every year, are especially severe in logged and degraded forests (fires often go out naturally in wet and dense old-growth tropical forests). Most of these fires in degraded areas are lit intentionally to clear and fertilize land cheaply for plantations. These can easily get out of control, however, sweeping across a vast terrain, including into peatlands and underground coal seams. During Indonesia's "fire season" a smoky haze routinely smothers the neighboring countries of Singapore and Malaysia. These fires alone are one of the largest sources of global greenhouse gas emissions. Some analysts now rank

Indonesia as the world's third largest greenhouse gas emitter – an estimate that the Indonesian government initially challenged, but later backed its inclusion in the 2009 Copenhagen negotiations on climate change (stressing that deforestation and forest degradation were accounting for up to 85 percent of its total emissions).[14]

Rising consumption of beef, meanwhile, is encouraging land-clearing for cattle ranches in Brazil (the fourth largest global greenhouse gas emitter), which over the last decade has taken over as the world's largest beef exporter by volume. Clearing land to plant soybeans for export – used for everything from animal feed to filler for processed foods – is another big cause of deforestation in Brazil. A fifth of the Brazilian Amazon is already gone. And every year another couple of million hectares are cut and burned down.

Given the global political economy of timber, what can be done to reduce the increasingly grave costs of timber consumption? How can we save the world's remaining old-growth forests? Can international agreements and financing help to control corporations and empower communities? What about voluntary certification and corporate social responsibility? Can these help to govern global commodity chains?

Governing timber consumption

Forests cover roughly 30 percent of the earth's land area. Governing commodity chains to reduce the rising environmental and social costs of producing and consuming timber from these diverse and extensive forests – or at least distribute them more equitably – is certainly one of, if not the most, difficult of all global environmental challenges. Indicative is the lack of an international environmental agreement on forest management despite decades of international meetings to consider establishing one.

Many factors complicate and impede the capacity to govern timber consumption on a global scale. A rising world population and increasing affluence are two of the most significant. During the second half of the twentieth century the global population grew more than it had in the preceding 4 million years. Since the 1970s the world economy has been growing even faster than population, nearly tripling in real terms from 1970 to 2000. It grew even faster from 2001 to 2006: quicker than during any five-year period since World War II.

The global financial downturn did slow world economic growth in 2008 and 2009. Yet the future will certainly see a much bigger global population and even higher per capita consumption. Most analysts still expect the global middle class to triple by 2030. And by the middle of this century the world population is set to exceed 9 billion, with more than 95 percent of this increase occurring in developing countries such as Indonesia, India, and China. In this setting, changing the choices of enough consumers fast enough to make a *global* difference for the world's forests is very hard – and getting harder with each passing day.

Many other forces and factors also impede the sustainability of timber consumption. One obvious one is advertising. Trillions of dollars are spent every year to convince people of the value of consuming various products, gadgets, and gizmos – and of the value of getting these at a bargain. This is deepening a culture of consumerism and reinforcing the growing discount-driven global economy. All of the messaging is encouraging consumers to buy more. Advertisers creatively promote "perceived obsolescence" and embed individual consumption with feelings like self-worth, freedom, adventure, and success. Getting a "deal" – saving hard-earned income – is a bonus for these feelings.

Advertising is just one of the many factors making it difficult to influence individual consumption with education or information. Availability and affordability, future savings, habits and convenience: all of these influence consumers, too. Many prioritize ease of purchase and good value; some are also inconsistent, saying one thing and doing another. Genuine and perceived uncertainty in what actually comprises a "sustainable purchase" or a "sustainable lifestyle" can also cause consumers to lose interest or confidence, justifying non-environmental choices. Consumers act, moreover, not only for personal reasons, but also in response to constraints, opportunities, and expectations: all of which the globalizing commodity chains shape and arguably even construct.

Recent trends within these chains are creating even greater challenges for state-level environmental regulators. At the same time, however, these trends are also opening up some innovative opportunities to improve environmental governance at the global level by leveraging the growing power of retailers to influence sustainable business practices through their global commodity chains. It is difficult, near impossible in many cases, for one state to govern a corporate timber chain comprising a loose group of affiliated firms scattered worldwide. Global advocacy groups face access, influence, and resource constraints. Governments and consumers commonly lack any knowledge of the original source or the full environmental and social costs of timber. For legal timber imports and exports, an increasingly globalized timber economy creates obscure chains of custody. For illegal production and trade, the task of monitoring, governing, and ensuring accountability is even tougher. Investigating the illegal activities within just one chain is not unlike a lone-wolf detective chasing thousands of suspects across hundreds of jurisdictions. False leads, dead ends, and incomplete evidence are unavoidable. And questioning all, or even most, of the suspects is virtually impossible, in part because the list of suspects is ever-changing in a corporate shell game of obfuscation.

Better corporate governance

Yet some trends present some opportunities to put in place new forms of "private environmental governance," such as industry-wide, voluntary eco-certification programs like forest certification and the sustainable palm oil and soy roundtable standards that can leverage global commodity chains to reach across jurisdictions, markets, and companies. Additionally, individual company-level global supply-chain greening initiatives – like product life-cycle and corporate footprint analysis, carbon accounting, supplier tracing, procurement requirements and audits, and sustainability performance tracking and reporting – can set new precedents for responsible corporate behavior, particularly in developing regions with weak regulation. In this sense, the increasing power of big retailers is opening new ways to influence corporate practices in states with weak governance capability, where state policies may look fine on paper, but where enforcement is poor or inconsistent (for many reasons, from low capacity to widespread corruption).

The reason is straightforward: companies must obey the specifications and standards of a big retailer like Walmart – or risk being replaced. Standard tactics that a small or large company might have employed in a weak regulatory state to evade government rules may no longer work as effectively. Big retailers, for example, are not open to bribes and care less about the economic stability of supplier regions. The risk for loggers and processors of violating a corporate rule is also much greater than for breaking a state rule. The penalty is not a government reprimand or fine, but the loss of a huge market (Walmart alone is a much bigger "market" than most states).

So far, however, as we have stressed in this opening chapter, the growing power of retailers over timber supply chains *is not* translating into better environmental practices on the ground. There is *potential* here. And many of the large

Northern retailers are now introducing sustainability policies and targets. Walmart, for example, announced in February 2010 a plan to reduce its global carbon footprint growth by 150 percent over the next five years (20 million tonnes) – 90 percent of which will come from requiring its more than 100,000 suppliers to change practices. But it is only part of the solution and, for the most part, the Home Depots and Walmarts of the world are still concentrating primarily on enforcing a very different set of rules: ones to ensure that suppliers meet design requirements and volume-targets, and, most importantly, keep costs as low as possible. The result is not better, but worse, environmental management, not just of timber but also the many products relying on and contributing to deforestation.

Going forward

The lesson here is important. States and consumers need to push retailers much harder to use their growing power more responsibly, so they become forces of environmentalism within globalizing commodity chains rather than forces of unsustainable consumerism. To date, however, global governance strategies to manage the world's forests are failing.

Imbalances in state governance over the last decade have created an inequality of progress toward more sustainable practices, one reflecting the inequality of global capitalism. Forest management is improving in strong regulatory states. Weaker states are seeing some improvements, too, especially where social networks develop strong independence from governments, as in the case of reforestation by the Green Belt Movement – a non-governmental global forest advocacy and reforestation initiative based in Kenya. Yet the overall trend is toward a deforested Third World growing and grazing cheap food, animal feed, fiber, and biofuel for export. No doubt this is putting short-term profits into some pockets. But in terms of long-term value, it is in no one's future interest: not environmentally, not socially, not economically.

One conclusion is clear. Transforming global commodity chains that underpin a world capitalist economy that undervalues natural capital is vital for global sustainability. Doing so will require *internal* adjustments within corporate commodity chains – such as chain of custody eco-certification, carbon accounting, and life-cycle assessments. But this is only part of the picture. Also essential are more systemic *external* changes to the industrial use, marketing, and valuation of forests through a convergence of governing forces that includes civil-society activism and strong regulation, as well as voluntary corporate social responsibility: the topic of our conclusion (chapter 6). Before assessing current efforts to govern global commodity chains, however, we need a more complete picture of how and why globalizing timber chains of big box retailers (chapter 2), First World forest and paper MNCs (chapter 3), and Third World companies (chapter 4) are causing deforestation.

CHAPTER TWO

The Power of Big Retail

Walmart topped the 2010 Global Fortune 500 list of the world's biggest companies – more than US$120 billion ahead of the revenues of the second and third place oil multinationals, Royal Dutch Shell and Exxon Mobil. Its US$400 billion or so in annual revenues tops the gross domestic product (GDP) of most states, including Sweden, Austria, Israel, and South Africa. If it were a national economy, it would rank as the 22nd largest in the world. Every week more than 170 million shoppers visit a Walmart store to buy thousands of different food, clothing, pharmaceutical, and electronic items. Customers can choose, too, from hundreds of timber products: from self-assembly tables, chairs, dressers, and bookcases to wrapping paper, notebooks, picture frames, hockey sticks, rolling pins, toilet paper, and diapers.

Wood or paper products can involve different corporate networks, production processes, and geographic regions. Yet all originate in a forest or timber plantation. And, as we started to explore in the last chapter, most follow similar lengthy global production pathways: traveling thousands of kilometers back and forth along commodity chains that criss-cross oceans and continents, linking companies headquartered in the North to cheap fiber and low-cost manufacturing in the South, especially in Asia.

Through these chains, big box retail stores in the North, such as Home Depot, IKEA, and Staples, are able to offer timber products at rock-bottom prices. Similar commodity chains are also providing these big box stores – as well as multinational hypermarket stores like Tesco, Metro, and Carrefour – with purchasing deals for reams of low-cost products made from palm oil, soy, and cattle: much of which, as chapter 5 will document in detail, grows on recently deforested land of the global South. While Southern forests deteriorate, big retail companies are prospering – rapidly growing in scale, market control, and global reach. The Global Fortune 500 list of the world's biggest companies did not even include "retail" as a category 20 years ago. Today the number of retailers on this list is edging toward 50.

Powerhouses in international markets, the influence of big retail within global commodity chains is steadily increasing as big box sales expand into emerging markets and as channels narrow toward a smaller and smaller number of giant discount stores. For timber, the growing buying power of these retailers is allowing them to shape more and more of the world's timber production, consumption, and trade. The world's top four timber retailers are Home Depot, Lowe's, IKEA, and Walmart. Within Europe, the leading timber retailer is Kingfisher, marketing under B&Q in the United Kingdom (UK) and Castorama and Brico Depot in France. Office supply stores such as Staples and Office Depot are also leading buyers and sellers of finished paper products.

Timber producers and manufacturers jostle for an opportunity to become part of a supply chain linked to a big box retailer. Many will redesign products, upgrade factories, and slash costs and profit margins to obtain – and then later meet – the high-volume orders of a multinational retailer. Once linked into the chain, however, many face ongoing pressure to continue cutting costs to keep prices as low as possible – in part because the supplier can no longer afford to lose the retailer's big-volume purchase order. Sometimes this pressure leads to innovation: for example, reducing energy and material usage. But, as we will show in chapter 4, more often it means that the supplier turns to questionable business practices, such as lowering product quality, cutting worker wages, benefits, and safety provisions, or logging and sourcing illegal wood.

This is especially true in regions with less governance capacity to control business practices. Increasingly, as we will analyze in detail in our concluding chapter, retailers are recognizing this as a potential risk and are starting to monitor and manage the impacts of their global commodity chains through supplier audits and green procurement. Yet, so far, as we document here in chapter 2 as well as chapters 3–5, a globalizing and growing discount economy of corporate commodity chains is driving deforestation of the world's tropical and boreal forests to supply increasing numbers of well-off, over-fed consumers with everything from throw-away paper to convenient flat-pack furniture to fast-food hamburgers. We begin in this chapter by analyzing the consequences of the rising power of Northern retailers within a world discount economy of globalizing commodity chains.

Retail and the discount economy

In the 1980s, business expert Michael Porter wrote in his books, *Competitive Strategy* and *Competitive Advantage*, of the three main strategies that companies were using to gain market advantage. These were cost leadership, product differentiation (e.g., branding), and niche marketing (e.g., segmentation).[1] In other words, successful corporations either had the lowest costs or produced a unique product.

This remains true for many companies. Yet, since the 1980s, a new corporate model has emerged at the core of the global economy. Some of the world's biggest companies are now achieving unprecedented economies of scale and market

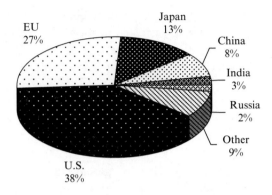

Source: Euromonitor International, *World Retail Data & Statistics,*
2008/2009 (London: Euromonitor Publications, 2008)

Figure 2.1 Percentage global retail sales by country, 2007

dominance through a global discount strategy involving the
intentional pursuit of very small profit margins on extremely
high volumes. These companies source low-priced raw materials from every region, squeeze offshore suppliers to lower
costs, ship products back and forth across and between continents, and set retail prices at close to wholesale cost. The goal
is the same as for any corporation: maximizing net profits.
But this is done by relying on huge sales turnover on small
profit margins. Retail companies following this model – most
famously Walmart – have achieved remarkable growth over
the last two decades.

Post-industrial countries depend on increasing rates of
consumption for economic prosperity. Retail now anchors
these service-focused economies. Consumer spending in
the developed world accounts for roughly 70 percent of all
employment and economic activity. Developed economies, as
figure 2.1 shows, dominate the global retail sector, accounting
for about 80 percent of world retail sales in 2007. This is starting to change as consumption increases quickly in emerging

economies like China, India, and Russia. Yet even these three countries still account for less than 15 percent of the global retail market.

International retail now exceeds US$13 trillion a year and accounts for almost one third of world GDP (9 percent directly and 20 percent indirectly). The central role that retailers are playing in expanding the world economy helps to explain why Walmart was able to rise so quickly over the last two decades. (By revenues, the top five on the 2010 Fortune 500 list of the world's largest companies were: Walmart, Royal Dutch Shell, Exxon Mobil, BP, and Toyota Motor.) This helps explain, too, why big box retailers are steadily gaining influence over global timber consumption.

Big box retail
A big box or "large-format" retailer is typically defined by its warehouse size (10,000 to 35,000 square meters), although stores in Europe and the UK tend to be smaller, with 5,000 square meters regarded as large format. Four features are common to most big box stores: no-frills, steel-framed box-shaped outlets with large adjacent parking lots; high-volume and small-margin business strategy; global low-cost product sourcing; and an emphasis on bargain-pricing. There are several different types of big box retailers (with many stores overlapping between categories), including:

- Hypermarkets (food plus non-food merchandise): Walmart, Tesco, Carrefour, Metro
- Discount department stores (non-food): Target, Kmart, Marks & Spencer
- Super food warehouses (food): Sainsbury, Kroger, Safeway
- Category specialists ("category killer"): Home Depot, IKEA, Staples
- Warehouse membership clubs: Costco, Sam's Club.

Big box retailers rose to global prominence in the 1990s, but formed out of a longer history of discount selling. Early discount companies were set up in the 1960s to take advantage of the post-World War II growth in suburban areas of the North. For example, the world's second largest retailer, Carrefour, was established in 1963 in the Paris suburb of Sainte-Geneviève-des-Bois. Walmart, Kmart, and Target were also established in the early 1960s on the outskirts of big American cities.

The second phase of discount retail was in the 1970s, marked by Sam Walton (Walmart founder) developing a warehouse distribution system and opening the first warehouse-style store in 1976. The store was "no-frills," selling products priced at close to the wholesale price. The warehouse format became popular in the 1980s and more and more consumers began to expect products at near cost. Bargain-hunting consumerism, along with more opportunities to leverage cheap offshore manufacturing (particularly in China), have contributed to a rapid rise of big box retail dominance since the 1990s, including over global timber consumption.[2]

Big box retail and timber consumption

Demand for natural resources is increasing as populations and economies grow, as industrialization expands, and as consumer incomes rise. Timber consumption is on track to increase by more than 50 percent by 2050 (from 1990 levels), with paper products alone expected to increase by more than 75 percent by 2020 (from 1995 levels). Already, global paper consumption more than tripled during the second half of the twentieth century. Demand for paper packaging, which accounts for half of paper production (by weight), is rising especially quickly as South–North trade and the online retail market ("e-tailing") expand. American online retail sales in

2009, for example, surged to 6 percent of total US retail sales. And over the next decade most analysts are expecting online retailers like Amazon.com and eBay to keep shifting global purchasing habits from "bricks" to "clicks." Walmart's site-to-store online shopping program (Walmart.com) hit a billion visitors in 2009.

Northern consumers account for the vast majority of this timber consumption. As we noted in chapter 1, developed countries, with just one fifth of the world's population, consume about three-quarters of the world's solid wood and almost two thirds of its paper. This consumption picture mirrors the overall retail sector, where six of the world's ten largest retail companies in 2010 were American and the other four were European.[3]

Big box retailers are among the world's largest timber buyers. This includes those specializing in home improvement, furnishings, and office supplies, as well as those combining supermarkets and department stores (the so-called hypermarkets). (This assessment is based on total revenue figures as disaggregated product sales data are confidentially held by the respective companies.)

As markets narrow toward big box stores, these growing companies are playing an increasingly influential role in shaping how and where timber products are made and traded. They are also shaping consumer preferences and behavior. Moreover, multinational grocery stores like Tesco, Metro, and Carrefour are indirectly influencing the global timber economy through the purchase and retailing of what some call "timber risk commodities," such as palm oil, soy, and beef – all of which, as we will document in chapter 5, are significant drivers of tropical deforestation.

Home Depot is the largest wood product retailer, followed by Lowe's, IKEA, and Walmart (table 2.1). Global office supply chains such as Staples and Office Depot (with 2009 revenues

Table 2.1 The world's biggest wood retailers		
Company	Retail Sales, 2009 ($US million)	Global scale
1. Home Depot	66,176	World's largest home improvement chain with more than 2200 stores mainly in the US as well as five other countries.
2. Lowe's	47,220	World's 2nd largest home improvement chain with 1700 stores in the US, Canada, and Mexico.
3. IKEA	32,475	Discount furniture chain pioneer founded in Sweden, with 301 stores (267 IKEA group + 34 franchisees) in 37 countries.
4. Walmart	408,214	World's largest retailer and the world's biggest company (in terms of revenues) with more than 8,000 stores in 15 countries.

Sources: Company Financial Reports

of more than US$24 billion and US$12 billion respectively) are also leading buyers and sellers of finished paper products. Nonetheless, in recent years the do-it-yourself home improvement sector has been the most important retail purchaser, growing before the financial crisis of 2007–09 on average at above 5 percent in the United States and 10–15 percent in Asia (especially in China, Japan, and India). Home Depot and Lowe's are the top do-it-yourself companies in the United States. Kingfisher (B&Q) is the largest one in Europe.

As table 2.1 summarizes, Home Depot maintains its position as the world's largest timber retailer through more than 2,200 stores in the United States, Canada, Mexico, and China, with 22 million customers going through every week. Lowe's holds on to second place through 1,700 stores across the US, Canada, and Mexico. The "pioneer discounter" IKEA, the world's largest furniture manufacturer, holds third place by selling more than 9,000 wood products across 35 countries in 267 retail warehouses – some up to 366,000 square feet in size, or seven times the size of an American football field. The 1.6 million customers that visit an IKEA store every day combine to consume more than 7 million cubic meters of wood per year. This is equal to one tenth of Sweden's annual forest production. Every year IKEA also uses more than 100,000 tonnes of paper to produce its catalogue. The company prints close to 200 million copies a year – 56 editions in 27 languages across 38 countries – making it one of the world's bigger paper consumers (see box 2.1).

Walmart, by selling more than 900 different forest and paper products, holds the fourth spot as the world's largest timber buyer. Yet its impact on deforestation is in many ways much greater than Home Depot, Lowe's, and IKEA, as it sells far more products that contain ingredients such as soy and palm oil. The company is now almost two times bigger than its seven closest competitors combined, with more than 8,400 stores in 15 countries. Not all of its stores run under the Walmart brand. In the UK, for example, Walmart operates under its wholly owned subsidiary, Asda – the second biggest grocery superstore in the country, behind Tesco.

Just over half of Walmart's retail outlets are in the United States – strategically spread out across the country so that more than 80 percent of Americans live within 24 kilometers (15 miles) of a Walmart store. In 2009, Walmart accounted for about 11 percent of Americans' total expenditures on retail goods. Its subsidiary Asda, meanwhile, accounted for 17 percent of the UK grocery market. On a global scale such a narrowing of consumer spending began to take off in the 1990s, with the accelerating growth of retail and with the rise in the market power of big box companies.[4]

Accelerated growth of retail

Retail did not become a business category within the Global Fortune 500 until 1995. By 2009, 47 of the Global Fortune 500 biggest companies in the world were retailers.

Walmart's rise has been meteoric (figure 2.2). Its revenues in 1979 were a respectable US$1 billion. Yet, by

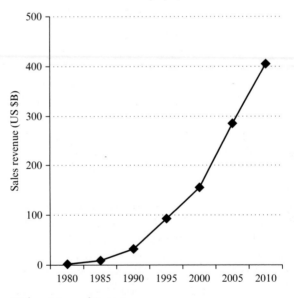

Source: Walmart Annual Reports

Figure 2.2 Walmart growth in revenues (1980–2009)

1994, these were more than US$1 billion *per week*. Since then Walmart's growth has been even more remarkable. As reported in Walmart's financial statements, revenue in 2005 was US$285 billion and the company had 1,500 stores. Just five years later sales were more than US$400 billion – and nearly 7,000 new stores had opened. At times, the pace of expansion has bordered on frenetic: on average from 2000 to 2007, a new Walmart store was opening in the US every day or so.

Other big box retailers have also been expanding rapidly over the last two decades. Lowe's annual reports show the company was completing new store projects in the US at a rate of more than one per week in the mid-1990s as the do-it-yourself home improvement sector boomed. The revenues of Home Depot also grew exponentially over the past two decades: from US$1 billion in 1986 to US$15.5 billion in 1995 to more than US$70 billion by 2008 (it slumped in 2009 with the stall in US home building and renovating during the global financial crisis, but looks well set to rebound once the American home improvement market recovers).

The American home improvement retailing sector has been booming since the 1980s. The 1997 *Lowe's Annual Report* shows this sector grew from US$55 billion in 1982 to US$142 billion in 1997. Today it is more than US$270 billion. The US Home Improvement Research Institute predicts it will keep expanding following the 2007–09 global financial downturn, likely exceeding US$335 billion by the year 2014. Globally, the US accounts for roughly half of the global home improvement market's value (47 percent in 2008). In 2008, this market had a value of US$570 billion; analysts at the Home Improvement Research Institute also expect the global market to rebound from the 2007–09 downturn, exceeding US$635 billion by 2013.

Logistics   technology   and   offshore   outsourcing   of

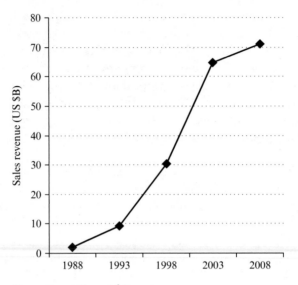

Source: Home Depot Financial Reports

Figure 2.3 Home Depot growth in revenues (1988–2008)

manufacturing to low-cost countries have been major factors propelling retail growth since the 1990s. China has been
especially important. The country doubled its share of the
world economy from 1990 to 2000, in the process becoming a leading exporter and lowering the retail prices of some
merchandise by as much as 90 percent.[5] The big retailers are
all sourcing in China. Walmart established its first procurement office in Shenzhen in 2002: its thousands of suppliers
in China now provide up to 80,000 different consumer items,
including many manufactured from wood and shipped and
branded with paper packaging. The trends to outsource manufacturing and procure goods in countries like China are two
of the key reasons retail markets are gradually narrowing
toward big box retailers.

Retail market concentration

Global markets for consumer goods, including for manufactured timber products, remain largely fragmented.
Nonetheless, consumer markets in the First World are gradually narrowing under the control of large discount retailers.
Walmart alone now accounts for about 15 percent of US
imports of consumer goods from China. Similar trends are
occurring for timber. Home Depot's orders from Sweden
in recent years, for example, equate to about one quarter of
all solid wood shipped from Sweden to the United States –
and one tenth of all European timber exports to the United
States.

The Wal-Mart Effect, the title of investigative journalist
Charles Fishman's 2006 book, has become a popular phrase
to describe the transforming social and economic effects of big
box corporations, including the power to drive local retailers
out of business or out of town. Some analysts call a discount
company within a particular industry segment a "category
killer" – such as Home Depot and Lowe's in home improvement, IKEA in home furnishing, and Staples and Office Depot
in office supply – because these effectively eliminate smaller
retail categories: for example, hardware, furniture, stationery, garden centers, and lumber yards. These big box retailers
do this by saturating markets with high volumes and setting
prices so low that it is very hard for smaller, local businesses
to compete and remain profitable.

The high-volume purchases of big box retailers can also
shut out small retailers by making it difficult for them to
obtain – or maintain – an adequate supply of some products.
"The Home Depot is wrecking our industry," complained
one small California-based lumberyard owner in the late
1990s. "Louisiana-Pacific [a large integrated timber company]
dropped me and every other independent wholesaler they

have been dealing with for the past 30 years for Home Depot . . . I couldn't buy a stick from them."[6]

Retail markets in emerging economies remain more fragmented than those in the North. These economies, however, may soon follow the trend in the North. Big box retailers are rapidly expanding into developing markets through direct sales as well as by partnering and buying up established retail chains. IKEA has opened eight retail outlets across China. Kingfisher (B&Q's owner) was one of the first entrants into China in the late 1990s. In 2006, Home Depot purchased China's largest home retailer, The Home Way. Walmart's growth in Brazil (to become the country's third largest retailer) was propelled by the acquisition in 2004–05 of Bompreço and Sonae – two of the country's leading retailers. In recent years roughly 80 percent of Walmart's US$100 billion international business has been through acquisitions. With sales growth flattening in the US market, the company has focused its growth strategy on global more than domestic expansion, particularly in the emerging markets of China, India, and Russia.

So far, although foreign sales in the global South are rising, they still account for a relatively small portion of total revenues for big box retail. Walmart's international sales, for example, contribute to about one quarter of its total revenue. In some cases big box retailers are struggling in the South to overcome cultural differences, gain local trust, and control markets. Kingfisher's B&Q exited Taiwan and is now closing its stores in China. Kingfisher executives said the company expanded too quickly after overestimating the demand among Chinese consumers for do-it-yourself retail – with more consumers preferring to bring in contractors than is the case in Europe or North America. Home Depot pulled out of Chile in 2001 after entering just three years earlier. The company's approach, as business professor Constanza Bianchi reveals in her analysis of the case, caused fundamental problems. Locals did not perceive Home Depot's efforts to include its Chilean partner Falabella in decisions as sufficient. Home Depot was also seen as too secretive within a domestic tradition to share information and maintain good relations with competing firms. The company seemed to misjudge the market potential as well – most Chileans at the time were working long hours with little time for individual home improvement.[7]

Such difficulties have not stopped big box retailers from slowly expanding sales in developing markets. After initially faltering in its attempt to introduce the disposable paper diaper to Chinese consumers just over a decade ago, Procter & Gamble's *Pampers* is now the number one selling diaper in China. Building on its highly successful multimillion-dollar marketing strategy to promote the sleep benefits of the disposable versus reusable cloth diaper, Procter & Gamble has plans to capture 1 billion more customers throughout the global South over the next five years.

Rising per capita consumption of discounted products in the North remains, however, the principal driver of the growing world market control of big box retailers. Global commodity chains with deep links into the South have been essential for sustaining this rising consumption in the North.

Global timber commodity chains

A commodity chain refers to the network of producers, manufacturers, distributors, retailers, and end-consumers that participate in the production, delivery, and sale of a particular product.[8] Simplified, it is the pathway that a product travels from its raw material origin to the retail shelf (as figure 2.4 shows for timber). In general, wood fiber passes from upstream loggers to primary manufacturers (e.g., sawmills and pulp mills that turn logs into pulp, lumber, and panels) to a range of secondary manufacturers who convert the pulp

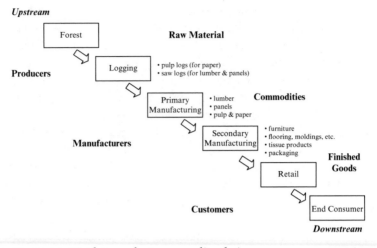

Figure 2.4 Timber product commodity chain

and manufactured solid wood into semi-finished and finished wood products (e.g., chairs, tables, moldings, cardboard
boxes, paper coffee cups, tissues, dinner napkins, and garden
stakes). Traders, distributors, and wholesalers facilitate the
movement of the goods downstream to customers, including
institutional buyers and retailers. Retail companies market
the timber products to the end-consumer (including businesses and individual shoppers).

Commodity chains were once fairly simple. For the most
part producers and consumers were located within the same
trading region and salespeople were in direct contact with
suppliers, writing up contracts to renew product orders.

These chains began to change fundamentally in the 1990s.
The rapid growth of export-oriented manufacturing in China
and other low-cost regions lengthened and twisted them
well beyond local and regional trade. The number of suppliers and the number of chains grew quickly, creating greater
distances and more obscure pathways between producers

and consumers. Also, the introduction of logistics technology – including universal product codes, radio frequency identification tags, and electronic data interchange – sped up the movement of goods and eliminated many middlemen (as well as the need for so much direct sales contact). (Universal product codes and radio frequency identification tags allow for the scanning of product information at the point of sale and electronic data interchange systems transmit information and orders electronically so that inventory can be updated immediately on a just-in-time basis.) Walmart's logistics technology for moving goods quickly across great distances has led the field, with even the Pentagon learning from the company's highly advanced computerized systems.

Commodity chains for timber share a similar overall structural flow from a forest through various stages of production to the end-consumer. Still, specific arrangements among corporations do vary within particular product chains. This affects, as the examples below show, such things as raw material requirements, the location of production, as well as the distribution of power and influence among various players.

Product chains
For wooden furniture, commodity chains are dispersed widely among many different contractual partners. The list is long and includes: loggers, log traders, sawmills, plywood and engineered wood facilities, and woodworking and furniture plants as well as furniture wholesalers, brokers, and distributors. Typically, downstream retailers wield significant design control and purchasing power within these chains. Specific design, material, and assembly requirements do tend to vary across categories of item: for example, across office, kitchen, bedroom, dining, and outdoor furniture. Also, unlike with a more standardized wood product such as flooring, a single mill cannot easily supply the variety of wood (e.g., particle

board and veneers) and different sizes of materials required for a product like furniture.

Furniture chains can vary significantly as well. Chains for specialized high-end pieces, for example, are usually quite different from ones for low-end, mass-produced "flat-pack" or "ready-to-assemble" ones. Smaller businesses tend to make more expensive, "crafted" furniture and are more likely to sell locally (shorter chains); on the other hand, big factories tend to manufacture the large-volume export orders for products for discount retailers (longer chains).[9] Increasingly, the global furniture industry is shifting toward factory production, especially self-assembly furnishings in Asia. One sign of this shift is the rise of Walmart as the largest seller of wooden furniture in the United States.

Unlike within the wood furniture sector, integrated multinational forest and paper companies are more prominent in commodity chains for manufactured solid wood and for paper and tissue products. These products tend to be more standardized and in the case of products such as tissue, are often branded. To control quality, volumes, and distribution, multinational companies often own many of the stages of production for these products. For example, Kimberly-Clark, the world's largest producer of tissue products, operates pulp and paper mills as well as converting plants to produce and brand goods – such as diapers and toilet paper – that the company then sells directly, in particular to big box retailers. Other global timber companies such as Weyerhaeuser, Stora Enso, and SCA are even more integrated – manufacturing and branding paper and solid wood products as well as sourcing fiber from their own forestlands (see chapter 3 for details).

Such control over long pieces of a chain, however, is starting to weaken somewhat. Multinational forest and paper producers certainly remain powerful within the global timber economy. But, increasingly, power over the functioning of commodity chains as a whole is diffusing toward retail companies, as the traditional timber multinationals divest a range of assets to focus on particular business segments (e.g., International Paper has been selling off its forestlands) and as consumer markets concentrate among the big box retailers.

Timber commodity chains not only vary by business segment and product, but also at the level of the individual firm. Each timber producer and retail buyer negotiates the arrangements for a particular supply chain. This creates some significant differences. Walmart, Home Depot, and Lowe's, for example, all essentially just provide retail shelf space (rather than manufacturing capacity). In contrast, IKEA designs its own flat-pack self-assembly furniture and, in 1991, formed the company Swedwood to oversee the manufacturing and supply of its specialized self-assembly furniture.

The distribution of value and profits also varies considerably across and within global commodity chains. Generally, local communities and small logging companies receive a fraction of the final profit. By the time it goes for further manufacturing in a place like China, the semi-finished timber product may be selling per cubic meter for 10–20 times the price paid for the log. The finished product may then sell to consumers in North America, Europe, Japan, or Australia – even through a big retail store – at 10 times as much again per cubic meter. Often "the people who benefit the least," as one study succinctly summarizes, "are local people whose forests are being exploited."[10]

Some practices are common among retailers despite distinct differences among particular supply chains. One of the significant trends is toward adopting the business practice of "disintermediation." This involves retail companies managing distribution within their networks by passing over brokers and wholesalers and buying directly from manufacturers. Some retailers are even going a step further: requiring suppliers to deliver merchandise to their stores directly to skip distribution altogether. According to Walmart, for example, 21 percent of its US purchases and 17 percent of its international segment purchases were shipped directly from suppliers to its retail outlets in 2009. Walmart's remaining purchases are handled through its vast network of distribution centers and huge fleet of trucks – with more trucks on US highways than any other company. (In addition to 129 distribution facilities in the United States, Walmart has another 132 across 14 countries.)

The reasons the big retailers are pursuing disintermediation are not hard to uncover. It enhances cost-efficiency, inventory management, and distribution speed. And, ultimately, it increases "buyer power" and market control.

Commodity chain buyer power

The rising buyer power of big box retailers is increasing the capacity of these companies to influence decisions about where forest products are made, what price they are sold at, how they are designed, and when they are shipped.[11] Contracts with a large retailer commonly require major investments in plant redesign. The scale of an order can even be so large that it takes up a considerable portion of the production capacity of a large integrated timber company.

Since 1997 the Swedish company SCA, for example, has shipped about one third of the annual production from its sawmill in Rundvik, Sweden, directly to Home Depot stores in the United States. Roughly every five weeks a ship leaves the Rundvik port for Philadelphia, where SCA then distributes the wood products by trucks and trains to Home Depot's regional warehouses. SCA also has a long-term agreement with IKEA (signed in 2009) to deliver every year more than 1 million pinewood "Gorm" shelving units. To meet this order, SCA needed to purchase new equipment and install and dedicate an entire new production line to scan, trim, nail, and flat-pack the shelving units. The IKEA order now accounts, according to SCA, for about 10 percent of the total capacity of SCA's sawmill in Bollsta, Sweden.

Other integrated timber MNCs are similarly starting to rely more on large orders from discount retailers. The Finnish-Swedish company Stora Enso, for example, relies on the annual paper order for the IKEA catalogue. More broadly, the company Procter & Gamble has 15 percent of its market tied to Walmart. This makes Walmart its biggest customer by far – with orders as large as its next nine customers combined. Proctor & Gamble now has a 250-person team located next to Walmart's head office in Bentonville just to maintain this relationship.

To offer consumers the promised "lowest prices," discount retailers are increasingly looking to link into the lowest-cost suppliers. Hundreds of thousands of companies in developing countries now count on sales to big retail companies. Walmart has 10,000 suppliers in China alone. It would be virtually impossible to document fully the great diversity of suppliers across chains, companies, and countries – plus these are continually changing and shifting as companies fold, reorganize, and negotiate new deals. Connections between suppliers and big box retailers can also be either direct or indirect. For example, a small mill in the western Brazilian state, Acre, has a direct one-to-one contract with the UK's B&Q to supply hundreds of thousands of tropical plywood sheets annually. But most of the relationships between Southern suppliers and Northern retailers are indirect: through a wholesaler or sub-contract arrangement with a larger producer (including Northern forest and paper companies) that contract-out to the smaller Southern supplier, essentially acting as a big box "agent," relaying the retailer's low-cost, high-volume demands.

Squeezing suppliers

It is not surprising that suppliers are attracted to linking into the commodity chain of a big box company. A local retailer may buy 20,000 units of, say, a wooden toy or picture frame. But a big box purchase may be in the range of 20 million units. Moreover, many suppliers (particularly in developing regions) are enticed by the chance to "upgrade" within the commodity chain.[12] This might mean achieving greater production efficiencies through new technologies, as well as the opportunity to "move up" from providing a lower-value raw material, such as sawlogs or wood chips, to producing a higher-value semi-finished product such as lumber, plywood, and paperboard, or perhaps an even higher-value finished timber product under the retailer's brand (e.g., self-assembly furniture, tool handles, decking, doors, and so on). Such opportunities, however, can come with considerable drawbacks. Because of the market clout of big box retail, within a short period of time suppliers (particularly smaller firms) can find themselves vulnerable – commissioned by the retail buyers to meet tough efficiency standards and low-cost requirements.

Squeezing suppliers is a critical factor allowing a big box retailer to offer "lower prices" and "more savings" than its competitors. Tactics include setting the initial contract price "low but reasonable," then adjusting prices downwards for future orders. A company like Walmart has so much purchasing power that at times it can simply tell suppliers what it is willing to pay rather than asking for a quote. Big box retailers have even been known to check the financial records of a supplier to see how much room it might have to cut its margins further – and thus lower its prices further. Large retailers also pressure suppliers to speed up the delivery of products; a supplier unable to tighten a delivery window may well get dropped.

Suppliers accept the retailer's terms and slash profit margins because they fear retaliation. Few suppliers of basic commodity items are even willing to criticize the big box retailers, as switching to a more compliant or less critical supplier is generally easy for these retailers because of the large number of eager substitute suppliers. The lack of disclosure about supplier–retailer deals and concessions also leaves big box retailers generally free to continue to use aggressive pressure tactics, gaining further market power.

Such pressure can improve operational efficiency of suppliers. Charles Fishman explains in the case of Walmart: "it is accepted wisdom that Walmart makes the companies it does business with more efficient and focused, leaner and faster. . ." Almost by necessity, suppliers become, "as microscopic as Walmart is at managing their own costs . . .. turning themselves into shadow versions of Walmart itself."[13] Yet, as the history of timber over the last two decades shows, the combined effect of this kind of pressure is also creating strong incentives for suppliers to cut corners and evade regulations, particularly in developing countries with low state capacity and inadequate oversight.

Controversial fiber sourcing

Manufacturers and timber producers find creative – and sometimes illegal – ways to lower prices for big retail buyers. For many in the South, as we will document in chapter 4, this has meant lowering employee wages and health-and-safety measures, purchasing more illegal timber, and adopting destructive forest practices. Manufacturers in developing countries that have redesigned their plants to acquire a big box order are particularly vulnerable when a retailer comes back year after year to demand a further lowering of costs. Many are unable to afford further retrofits having already cut back their margins and gone into debt. Desperate, they look to lower their input costs by passing the retailer's discount price demands up the supply chain to even more vulnerable players – often those logging the forests – ultimately creating incentives to avoid paying government fees, and for destructive harvesting and the smuggling of logs from parks and legal concessions.

For developing states and communities, the financial losses, let alone the social or environmental ones, run into billions of dollars. The World Bank, in a 2006 report on strengthening forest law and enforcement in developing countries, estimated that illegal logging on public lands was costing developing countries US$10 billion annually – plus another US$5 billion lost from evaded tax and royalty payments. Seneca Creek Associates, in a 2004 study, estimated that illegal timber worth about US$5 billion was now part of the global timber trade.

Much of this illegal timber ends up in manufactured wood and paper products on retail shelves in the First World. The case of China is revealing. About half of all of the world's traded timber, as we saw in chapter 1, is now passing through China. China is also using mountains of paper to package its growing exports of consumer goods. Meanwhile, in recent years 30–80 percent of timber imports into China have come from illegal sources: such as the primary forests of Russia, Indonesia, Malaysia, Thailand, Papua New Guinea, and Burma.

At the same time, big box retailers are relying more and more on sourcing low-cost semi-finished and finished wood and paper products from Chinese manufacturers. It seems reasonable to assume that at least some illegal timber must be making its way onto big box retail shelves, even if the lack of tracking through long and complex commodity chains makes it impossible to know precisely how much. The US Environmental Investigation Agency's (EIA) 2007 analysis of

Walmart would seem to support this. The EIA, analyzing data from US Customs from May 2006 to April 2007, found that 84 percent of Walmart's wood product import transactions were with China – noting further that many of these products were highly susceptible to containing illegally sourced timber, especially from the Russian Far East.

Imports from China are, of course, not the only ones likely to contain illegal wood. Manufactured timber products from neighboring countries such as Vietnam, Indonesia, Malaysia, and Thailand, for example, are similarly suspect given the high percentage of illegally sourced wood in these countries. Nor is Walmart the only – or necessarily the most likely – retailer to sell such products. A 2006 EIA report, for example, found that several big retailers, including Home Depot and Lowe's, were inadvertently sourcing illegal merbau flooring from Papua province in Indonesia. The retailers seemed unaware of the origin of the wood product.[14] In many cases this is not surprising, as suppliers to the big retailers often deal with many very small suppliers who lack any sort of computer tracking technology to allow anyone to check and verify the origin of particular products.

Big discount stores have also been under the spotlight for stocking wood products from high-biodiversity, controversial, and endangered forests. Advocacy groups, for example, have been targeting Home Depot for carrying cedar from Canada's west-coast rainforest; redwood from California; mahogany doors from the Brazilian Amazon; and luan plywood, wheelbarrow handles, and doors, as well as ramin dowels and tool handles from Southeast Asia. The European company Carrefour has also been under the spotlight for selling large amounts of uncertified wooden furniture from places such as Vietnam and Malaysia, although in recent years it has been trying to source more certified wood products.

Conclusion

The last two decades, then, have seen a striking rise within the corporate world of big box retailers. This trend looks set to continue for at least another half-century as global consumption is sure to keep growing along with the world population and national economies. Walmart now sits at the top. But its success means that it will more than likely face increasingly stiff competition for the established markets of the North and the emerging markets of the South. Already, dozens of other retail companies are now rising up the Fortune 500 list of the world's biggest companies – and many more are striving to join them.

The resulting increases in the consumption of low-priced products will certainly add even more pressures on the world's natural systems. Forests in the global South are especially vulnerable to this growing world discount economy. Consumption of cheap wood, paper, and packaging are all rising quickly – pushed along in the case of packaging by exploding world exports, and in emerging online retailing by companies such as Amazon.com as well as big box retailers themselves employing multichannel "bricks and clicks" sales strategies. Consumption of food and fuel products relying on deforested land in the South is rising, too. Such trends seem likely to continue even as retailers like Walmart start to exert more control over the social and environmental practices of its global commodity chains: a theme we will return to in our concluding chapter.

Big box retailers are partnering with more and more suppliers in the global South – particularly in China – to gain access to new markets, inexpensive facilities, and cheap inventory. These retailers have significant capacity to influence timber suppliers: potentially far more than a government in a developing economy, especially when the supplier is relying on the purchasing contract to pay for earlier upgrades or expansion necessary to get the contract in the first place. All of the big box retailers are now developing sustainability criteria and policies to raise supplier standards, although, so far, they have been far more likely to throw their weight around to get better financial deals.

Quality and design of products do matter for big box retailers. But their main focus is on getting large volumes of discounted items to consumers, fast. Thousands of wood products are now trading back and forth from South to North and North to South, often crossing borders many times, creating longer and more complicated chains from the beginning of the production process to the point of final consumption. Through these chains, profit margins per timber product or transaction are generally low – especially as the channels to end-consumers continue to narrow – with high overall profits for the big retailers. Southern suppliers, eager to gain global market access, and under pressure from big retailers to fill orders quickly and to keep costs as low as possible, turn to sourcing illegal wood, evading taxes, and logging destructively.

Big retailers, as this chapter highlights, are gaining more and more authority and influence over these suppliers within the global commodity chains. The next chapter shows, too, that the historic power of the Northern multinational wood and paper companies is diffusing on a global scale. Still, with giant facilities and billions in revenues these companies remain very strong in many regions and over segments of many chains. Analyzing in greater depth the role of these companies within the global commodity chains is therefore essential for a full understanding of past – and likely future – trends in the political economy of timber.

CHAPTER THREE

The Northern Forest and Paper Multinationals

Over the last century, Northern forest and paper multinational companies have innovated, vertically integrated, and pursued low-cost production. These strategies have helped companies such as America's International Paper and Sweden's SCA to emerge – and remain – as global powerhouses within the world timber industry. The last few decades, however, have seen significant strategic global corporate expansions within an overall, systemic shifting of the world economy toward emerging markets.

With sales and profits tightening in domestic markets, these companies have been working to maintain global market dominance by merging, partnering, and relocating to fast-growing markets in the South. Expanding Southern industrial timber production and marketing are driving and reinforcing the increasing per capita consumption of forest and paper products in these emerging markets. Rising timber production in the Third World is also driving up timber exports from the South to the North, contributing to an overall shift in the world timber economy from intraregional (within continents) to more interregional trade (between continents).

Such trends, we argue, are allowing Northern multinational corporations to maintain – and at times increase – their control over sections of global timber chains. Yet, as these globalizing chains lengthen to involve more low-cost offshore producers and manufacturers, the control of the traditional timber multinationals is also diffusing, particularly for more finished consumer timber products where, as we saw in chapter 2, market channels are increasingly concentrating toward big box retailers.

We begin our analysis of these shifting economies, business strategies, and power dynamics among Northern MNCs fighting to gain control of globalizing markets by unpacking further the language and statistics of commercial timber and the global forest and paper industry.

Commercial timber

The commodification of forests frames the geopolitics of timber. Of the 4 billion hectares of forest, about half is classified for wood and non-wood production and multiple uses (which includes logging, recreation, and preservation). Commercially, though, forests primarily supply wood. While non-wood forest products (NWFP) – such as food, medicinal and aromatic plants, tree roots, feed for livestock, and animal products – are also sometimes economically important, especially for local communities and subsistence farmers, these markets are smaller, often not priced, and largely unreported. Similarly, subsistence woodfuel consumption is significant in developing regions, but this is largely non-industrial, non-market, and local. The major contests over control of forest resources are tied to global markets and revolve around commercial timber – that is, around the 5,000 or so consumer wood and paper products that contribute more than US$1 trillion to a world economy of some US$60 trillion.

Commercial timber (industrial roundwood) is processed within two broad categories: solid wood and pulp. These are then converted into many different wood and paper products. As figure 3.1 shows, by volume sawlogs account for about 60 percent of global timber production and pulplogs for about 30 percent (about 10 percent is "wood residue" to produce

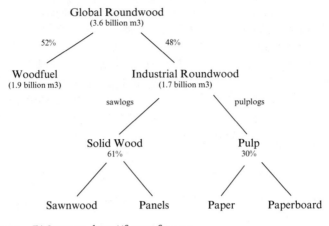

Source: FAOstat, at <http://faostat.fao.org>

Figure 3.1 Global timber product flow, 2007 (by volume)

products such as pellets). Sawlogs tend to be the bigger, higher-quality trees. These are processed into solid wood products, such as sawnwood (e.g., lumber) and panels (e.g., plywood and veneer). Pulpwood logs tend to be lower-quality, smaller trees (including fast-growing plantation trees). These are ground up or chemically treated to produce wood pulp to make paper (e.g., newsprint and office paper), tissue products (e.g., diapers and toilet paper), and paperboard products (e.g., packaging). Recycled wastepaper and "leftovers" like branches and wood chips from sawmills are also major wood fiber feedstock for pulp mills. Just under half of the total global supply of pulp ends up going into the production of paper packaging. Most solid wood, on the other hand, becomes lumber (sawnwood) for construction.

Since the 1950s, the amount of wood recovered from a log has roughly doubled. As well, more of the forest has gained commercial timber value. Timber producers can now utilize virtually every tree in the forest, including poorer-quality ones as well as more species. In the pulp sector, for example, mills in the past almost exclusively used softwood trees like spruce, pine, and fir instead of hardwoods like aspen and birch because the longer softwood fibers produced a stronger paper product. However, advances in pulp and papermaking technologies have changed the market, making hardwood (including tropical timber) a valuable feedstock, especially for its properties in making smooth printing and writing paper. Eucalyptus pulp in particular is now sought after because it produces "soft" and "bright" tissue products – like toilet paper. As a consequence, more eucalyptus pulp (mainly from Latin America, China, and Indonesia) is now produced than all other hardwood pulp grades combined. The global trade of hardwood pulp has taken off as demand in the North soars; in 2005, hardwood market pulp overtook softwood market pulp for the first time. Overall, though, global softwood consumption across all timber product categories is still roughly double that of hardwood.

Although less timber goes into making paper than solid wood products, paper is a higher-market value commodity. By dollar value, solid wood constitutes around 40 percent of the global timber market while pulp and paper accounts for 60 percent. Similar to other resource sectors, the market value of timber increases with the level of processing. Wastepaper, pulp, and raw logs and chips are lower-value markets as compared to manufactured sawnwood, panels, paper, specialty-converted paper products (e.g., stationery, gift wrap, egg cartons, and decorative table napkins), and finished wood products (e.g., flooring, furniture, picture frames, chopsticks, and wooden coat hangers) (see figure 3.2). The economic value and profits from commercial timber tend to increase as wood fiber moves downstream from the forest to the end-retailer. The biggest timber companies in the world have therefore all integrated higher-value timber processing,

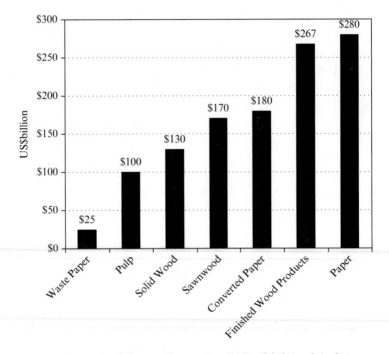

Source: International Finance Corporation (IFC), *Global Trends in the Forest Products Market, 2008,* Presentation by Richard English to India Farm Forestry Workshop (IFC New Delhi, 2008)

Figure 3.2 Global timber market size, 2007

manufacturing, and increasingly product branding into their business operations to remain profitable in a highly competitive global industry.

MNCs in the global forest and paper industry

Forestry is a capital-intensive industry with high barriers to entry. Modern pulp and paper mills, for example, typically cost more than US$1 billion. Moreover, because the industry in the North is mature, to continue to expand, the timber MNCs must compete hard with each other for sales – or diversify. The industry is also highly cyclical, closely tracking the ups and downs of the world economy. Bullish periods of strong economic growth see high demand for lumber and panels for building renovations and construction. Demand during these periods increases as well for wooden pallets and paper packaging to transport products worldwide. Consumption of newsprint, coated magazine paper, furniture, and books tends to go up as well. In recessionary times, demand for these types of products tends to falter, profits plunge, and mills close. Thus, chronic demand–supply imbalances and intense competition characterize the global forest industry.

Here, a small number of increasingly large MNCs jockey for optimal positioning to capture the greatest profits during upswings, while protecting themselves from the inevitable losses during downturns. As economic growth has been strongest lately in China, India, Russia, and Latin America, the most intense industry competition and greatest expansion have been occurring in these emerging markets. Thomas Gestrich, the president of International Paper in Asia, explains:

> 70 percent of the economic growth in the world comes from Asia and 40 percent of that from China in paperboard and pulp. So that is where the market is growing. We need to position ourselves to be part of the growth. We try to position ourselves in the right place at the right time.[1]

Concentrating and growing

Over the past 100 years, through rapid industrialization, followed by accelerating globalization and major consolidation, the forest industry has grown bigger and more concentrated. Fewer and bigger MNCs now control much of the world's timber production, trade, and consumption. Excluding private forest companies such as Georgia-Pacific and Asia Pulp and

Paper, the combined revenues of the hundred largest global forest and paper companies in 2008 totaled US$357 billion (just under 0.05 percent of the US$60 trillion world economy). The top 20 accounted for almost 60 percent of total industry sales.[2]

These companies have annual sales in the billions of dollars and workforces in the tens of thousands. Worldwide, they own sweeping plantations, hold licenses to log vast tracts of primary forest, and operate huge manufacturing facilities. For example, the International Council of Forest and Paper Associations (ICFPA), which represents the world's largest forestry companies, accounts globally for more than 90 percent of paper production and 50 percent of wood production.

Many small- and medium-sized companies also operate in the forestry sector. This is particularly the case in segments of the industry where advanced technology and large capital investments are less essential: for example, logging, sawmilling, and plywood manufacturing, as well as finished goods such as furniture, flooring, and frames. Still, the top MNCs tend to dominate the global forest and paper markets because of better access to financing for large-scale modern facilities and a bigger pool of highly skilled employees. For example, a single company, International Paper, now accounts for around 7 percent of the industry's total worldwide sales each year – an amount greater than the gross domestic product of many states. (Total company sales are not, of course, equal to a country's GDP. Still, comparing International Paper's sales to the World Bank's 2009 GDP figures is interesting, as the company would rank 85th out of 186 nations.)

Northern dominance
Over the last decade, as we discuss in detail in the next chapter, Third World MNCs from Brazil, Chile, Indonesia, Thailand, China, India, and Russia have been moving up the ranks of

Table 3.1 Top 15 largest global forest and paper companies, 2008		
Company name	Total sales/yr (US$ billions)	Head office
1. International Paper	24.8	US
2. Kimberly-Clark	19.4	US
3. SCA	16.9	Sweden
4. Stora Enso	16.2	Finland/Sweden
5. UPM-Kymmene	13.9	Finland
6. Oji Paper	12.8	Japan
7. Nippon Unipac	11.8	Japan
8. Smurfit Kappa	10.4	Ireland
9. Metsälitto	9.5	Finland
10. Mondi Group	9.3	UK/South Africa
11. Sequana Capital	7.3	France
12. Smurfit-Stone	7.0	US
13. AbitibiBowater	6.8	Canada
14. MeadWestvaco	6.6	US
15. Weyerhaeuser	6.6	US

Source: PricewaterhouseCoopers (PwC), *Global Forest, Paper & Packaging Industry Survey: 2009 Edition* (Vancouver, BC: PricewaterhouseCoopers LLP, 2009), 12

the world's top timber companies. Yet, in terms of sales, First World MNCs still dominate, with no Third World company in the top 15 publicly listed firms in 2008 (with the exception of Mondi, which is a jointly owned UK/South African company). Five of the top firms that year were American, three were Finnish, and two were Japanese (see table 3.1). Of the top 100 companies, more than three-quarters have their head offices in North America, Western Europe, or Japan; together these three places account for close to 90 percent of global forest and paper sales (table 3.2). (This ranking excludes several

Table 3.2	Global forest, paper, and packaging sales by country/region, 2008		
Country/region	Sales (US$ billion)	# Companies in top 100	% of Global sales
W. Europe	$128	30	36%
Finland	$44	5	12%
Sweden	$27	6	8%
US	$104	24	29%
Japan	$50	12	14%
Canada	$30	11	8%
Latin America	$18	9	5%
Emerging Asia*	$14	11	4%
China	$8	6	2%
India	$0.7	1	<1%
Russia	–	0	–
South Africa	$7	2	2%
Australia	$6	1	2%
TOTAL	$357	100	

*Note: Emerging Asian countries include: China, India, Thailand, Taiwan, and Korea.

Source: PricewaterhouseCoopers, Global Forest, 12–23

large private companies such as Georgia-Pacific, however, because the financial information is not publicly available. See note 2 for a listing of the major private timber firms.)

To gain and maintain such sales, First World timber MNCs are highly mobile, taking advantage of varying regional economic conditions and local resources to pursue joint ventures, private equity partnerships, and/or fully owned subsidiary operations around the world. International Paper, for example, has pulp, paper, and packaging facilities on five continents. For pulp, the company strategically produces at plant locations in five countries:

- Bleached eucalyptus kraft: Mogi Guaco & Luiz Antonio, Brazil
- Bleached chemi-thermomechanical pulp: Svetogorsk, Russia
- Northern bleached softwood and hardwood kraft: Kwidzyn, Poland
- Southern bleached softwood and hardwood kraft: North and South Carolina, United States, and Saillat, France.

All of the leading industry players have a similar focus on global expansion. Sweden's SCA manufactures packaging in 200 facilities across 28 countries. America's Kimberly-Clark produces diapers in 20 countries. The Finnish-Swedish company Stora Enso has paper mills in Europe, Brazil, China, and Russia. And America's Weyerhaeuser produces hardwood and softwood products in North America, Asia, Europe, and South America. Northern MNCs continue to control much of the global production even as facilities shift overseas, with ownership of the mills, for instance, still concentrated within these companies. Among the top 100 forest and paper companies, for example, North American and European multinationals still account for more than 60 percent of pulp production and more than 70 percent of paper production (by tonnage; see table 3.3).

Vertical integration
Leading forest companies – including International Paper, Georgia-Pacific, Weyerhaeuser, Kimberly-Clark, Stora Enso, SCA, UPM-Kymmene, and Oji Paper – all formed in the late nineteenth and early twentieth centuries. Many started out, such as Weyerhaeuser (see box 3.1), as small family-run businesses with a single mill and small forestlands.

Over the past 100 years, all of the leading global forest companies followed similar rapid growth trajectories that included

Table 3.3	Geographic concentration of pulp and paper production, 2007		
	Number of top 100 companies	% of market pulp production	% of paper and board production
North America	32	37.4	35.1
Europe	32	23.3	35.4
Asia	23	5.1	24.2
Latin America	9	31.4	2.9
Africa	2	3.0	2.1
Oceania	2	0.0	0.4

Source: RISI, *The PPI Top 100: Pulp and Paper International* (Boston, MA: RISI, 2009)

vertically integrating, diversifying, and expanding into overseas markets. Vertical integration involved establishing company operations to integrate all aspects of timber production and forest product trade: harvesting trees, producing and marketing lumber and pulp, distributing paper and packaging, and retailing branded consumer products such as tissues, paper cups, and diapers. Many forest companies also diversified into such businesses as real estate, mortgage banking, shipping, energy, chemicals, pharmaceuticals, and tourism.

Privileged timber access

Governments supported the growth of the large Northern forest companies. These companies, typically locating in less developed regions where timber was plentiful, have often acted as social agents. In many places they helped to set up basic community infrastructure, such as power lines and roads, as well as generating jobs and revenue for local, state, and national economies. A few towns – such as Kimberly, Wisconsin – are even named after the local forest company

Box 3.1 Weyerhaeuser Company: from small roots to global MNC

Fredrich Weyerhäuser, along with several other investors, founded Weyerhaeuser in 1900 when he bought 350,000 hectares of forestland in Washington State. At the time it was the largest private land deal in US history, and many thought it was a risky venture. Timber markets were thousands of kilometers away on the US east coast; a wild fire had recently destroyed neighboring forests; and the forests purchased were largely hemlock, which at the time had little commercial value as pulping technologies could not yet handle the qualities of the wood.

Despite the risks, the company under Fredrich's leadership continued to acquire land in the Pacific Northwest and US South, soon becoming the largest private forestland owner in the United States. It expanded into sawmilling and lumber wholesale retailing, as well as pulp and plywood production. The company expanded, too, into manufacturing products such as wooden shipping containers, grocery bags, milk cartons, corrugated packaging, and disposable diapers.

As the company passed from generation to generation of Weyerhaeuser offspring (e.g., Fredrich's son John was CEO from the 1920s to the 1950s while his great grandson George Jr. was president from 1966 to 1988), its operations grew and expanded overseas to the Philippines, Malaysia, Indonesia, and South America. The company also diversified into other businesses, such as chemicals, real estate, construction, mortgage lending, garden supply, and salmon farming (in the warm effluent from the pulp mills). The core business, however, remained lumber and building products and eventually the company divested most of these other businesses (as well as its forest holding in Southeast Asia).

Over this time Weyerhaeuser was transformed from a small family firm in Tacoma, Washington into the world's largest owner of softwood timber and the world's largest producer of softwood lumber and market pulp. The company also became a global leader in the production of cardboard, packaging, and disposable diapers. In 2007, prior to its fine paper divestiture to Domtar Inc., and the sale of its containerboard and packaging business, Weyerhaeuser had sales of US$16 billion, employed nearly 37,900 people, operated in 13 countries, and managed nine million hectares of forest in the US, Canada, Uruguay, and China. Since then, along with the global forest industry, the company has fallen on some hard times, with the 2010 Fortune 500 listing its revenues for the previous year at US$5.5 billion, and with its ranking among America's top companies falling from 236th in 2009 to 379th place in 2010.

Sources: Weyerhaeuser Annual Reports; J. A. Lamberg, J. Näsi, J. Ojala, and P. Sajasalo, eds., *The Evolution of Competitive Strategies in Global Forestry Industries: Comparative Perspectives* (Dordrecht, The Netherlands: Springer, 2006), 81–7; 2010 Fortune 500 Ranking of America's Largest Corporations, snapshot of Weyerhaeuser, at <http://money.cnn.com/magazines/fortune/fortune500/2010/snapshots/443.html>

(Kimberly-Clark). Commonly, as a timber company's economic power grew so too did its influence over communities and governments. In much of the world this has translated into privileged corporate access to forests, as well as government subsidies and incentives to support industrial forestry activities.

Early in the last century, forest companies acquired control over vast areas of North American and European timber. They did this by purchasing private forestlands, contracting with small family forest owners, and gaining access to trees on public lands through forest concessions (a lease to a private operator to harvest timber in a defined area for a period of time). During this period Northern forest companies gained competitive advantages in two main ways: by accessing local forests and by improving pulping or wood-processing techniques to use timber more efficiently.

Things began to change in the 1950s. Innovation and access to local fiber remained important. Yet, as the industry matured and inexpensive local timber supplies dwindled, the scale and scope of operations became the crucial factors for profits and growth. This led to mergers, acquisitions, and overseas expansions.[3] More recently, many companies have also been specializing more – outsourcing and contracting across geographic regions rather than vertically integrating within particular markets (see table 3.4).[4] One of the core goals of MNCs working within these global commodity chains is to create ways of maintaining low-cost production.

Maintaining the Northern competitive advantage

Consolidation and specialization

Timber, like automobiles and petroleum, is a mature industry in Northern markets with decreasing returns and an overall

Table 3.4	Global operations of the top multinational forest companies, 2009		
Company	**Created**	**Location**	**Global scale**
International Paper	1898	US	Biggest forest company in the world. 200 facilities on five continents.
Weyerhaeuser	1900	US	World's largest owner of softwood timber and largest global producer of softwood lumber and market pulp.
Kimberly-Clark	1872	US	World's largest producer of personal care tissue paper products. Product sales in 150 countries and manufacturing facilities in 41 countries.
Stora Enso	1918	Finland/Sweden	Fully integrated global forest and paper company with 85 production facilities in more than 35 countries. Enso (Finland) merged with Stora (Sweden) in 1998. Owns 2.1 million hectares of forestland in the Nordic countries and tree plantations in Brazil, Uruguay, China, Laos, and Thailand.
SCA	1929	Sweden	Integrated global forest and paper company with more than 200 facilities in more than 90 countries. Europe's largest producer of corrugated containers and personal care tissue products and largest private forestland owner (with 2.6 million hectares in northern Sweden).
UPM-Kymmene	1920	Finland	World's largest producer of magazine paper with 100 production facilities in 14 countries and owner of approximately 1 million hectares of forest in Finland, the US, and Great Britain, as well as plantations in Uruguay. UPM merged with Kymmene in 1995.
Oji Paper	1873	Japan	Japan's second largest papermaker, with 86 production sites in Japan, and forestry operations in Australia, Canada, China, Germany, and Brazil as well as plantations in China, Vietnam, Australia, and New Zealand.

growth rate that falls short of world economic expansion. This in part explains why companies seek competitive gains by strategically investing, divesting, and diversifying their assets as well as pursuing opportunities for low-cost expansion. Cost reduction is particularly critical: like other commodity businesses, timber products compete mainly on price rather than on design.

Consolidation is a key strategy here as it increases opportunities for economies of scale and production efficiencies. Mergers allow facilities to expand, technologies to advance, production volumes to increase, and ultimately per unit production costs to decrease. Also, market concentration promotes price stability, an asset in an industry that regularly faces wide price swings.

Consolidation started after the end of World War II, peaked in the 1990s, and continues today. In 2007 alone, for example, forest companies made 360 major deals worth US$27.6 billion. This included the US$4.2 billion merger of Abitibi-Consolidated with Bowater to create the world's largest newsprint producer. And it included International Paper's US$6 billion purchase of Weyerhaeuser's containerboard business – a move that made International Paper the largest paper-packaging producer in North America.[5]

More recently, as the industry continues to globalize, the big Northern MNCs are increasingly specializing within certain business segments and forming partnerships – outsourcing and contracting with other firms rather than vertically integrating production within a particular domestic market. Virtually all of the American-owned timber MNCs, for example, have sold off large areas of their domestic forestland holdings to infuse cash, reduce risk, and get better financial returns from buying fiber on the open market. International Paper has not only divested its timberland holdings, but has also sold a range of its solidwood and paper-manufacturing facilities to increase its focus on containerboard production.

Low-cost production

Keeping fiber prices as low as possible is vital for the profitability of timber MNCs in today's highly competitive forest industry. The cost of fiber can account for as much as 60 percent of production costs. Wood is roughly two times cheaper to produce in a tropical region like the Brazilian Amazon than in a temperate/boreal one like Sweden. Also, margins in emerging markets can be more than double those in the developed ones (mainly due to lower operating costs, less regulation, and faster-growing trees). Thus, over the last few decades, all of the global forest companies have been pursuing interests in Southern timber (particularly industrial timber plantations) and manufacturing facilities in the global South.[6]

Securing a low-cost fiber supply has been a core competitive driver of the expanding global forest industry. In order to offset fiber shortfalls, over the last century Northern MNCs have shifted investment from region to region in a quest for untapped primary forests. Historically, Japanese trading companies such as Mitsubishi Corporation and Sumitomo Corporation had timber interests throughout Southeast Asia, in particular the forests of the Philippines (1950s–1960s), Indonesia and Malaysia (1970s–1990s), and Papua New Guinea (1980s–1990s). European corporations historically dominated investments in logging and timber industries in Africa within the former European colonies, such as the Democratic Republic of Congo, Cameroon, Central African Republic, Gabon, Ivory Coast, and Liberia. And North American timber companies, such as Weyerhaeuser and International Paper, not only controlled vast quantities of American and Canadian timber but also held concessions and logging operations overseas in places like Southeast Asia (e.g., Indonesia and the Philippines in the 1960s and 1970s), as well as in South America.

Over the last few decades, foreign companies expanded in

particular into richly forested developing countries, such as Indonesia, Malaysia, Papua New Guinea, Cambodia, Burma, the Democratic Republic of Congo, Brazil, and Russia. In many cases, though, it is local loggers – not the Northern MNCs – who are now carrying out the harvesting. Following the first waves of logging, "second cuts" are now increasingly common in the tropics (particularly in Southeast Asia). On the first cut, typically the highest-value commercial species were selectively harvested: what is sometimes called high-grading. On the "second cut," loggers are now extracting as much of the lower-value fiber as is economically possible to supply pulp mills and solid wood manufacturing facilities. On occasion, loggers will now even clear-cut these degraded forests rather than selectively harvest specific species or trees.

The destruction of Southern forests has not gone unnoticed and Northern companies are now increasingly concerned with the high potential business risks of sourcing illegal timber from unsustainably managed native tropical forests. To replace supplies and lower risk, these companies are investing more and more in fast-growing tree plantations in South America, Central and Southern Africa, Asia, and Oceania. One common strategy is to develop a plantation alongside a pulp and paper facility – with larger plantations next to bigger mills. Development agencies like the World Bank and its member group the International Finance Corporation have encouraged the corporatization of Southern forests by providing subsidies and loans for industrial timber plantation development. Governments in these developing regions have in turn provided foreign investors with access to public land for these plantations to entice them with an economical, reliable, and consistent fiber supply.[7] Just a few MNCs can end up with a big share of a country's plantations. In Uruguay, for example, in 2010 just four foreign-owned MNCs controlled 40 percent of the industrial eucalyptus, acacia, and pine plantations: Botnia (Finland), Ence (Spain), Stora Enso (Finland-Sweden), and Weyerhaeuser (US). These four companies rely on these plantations to feed their giant pulp mills.

The case of Veracel

The story in Brazil of the pulp and paper company, Veracel, is typical of how these Northern corporate arrangements now operate in the South. Veracel is located in the state of Bahia, Brazil. It operates one of the world's largest single-line eucalyptus pulp facilities, supported by more than 90,000 hectares of eucalyptus plantation. The company is a joint-venture partnership of the Norwegian-Brazilian company, Aracruz Celulose (now called Fibria), the largest global producer of bleached eucalyptus pulp; and of the Finnish-Swedish company, Stora Enso, an integrated global producer of paper, packaging, and forest products.

Veracel's US$1.25 billion pulp mill – Brazil's largest private investment to date – opened in 2005. Directly employing around 400 people, the investment equated to roughly US$3.15 million per job. The mill, with a capacity to produce 1 million tonnes per year, set two new world records in its first year – producing almost 3,800 tonnes a day of high-quality eucalyptus pulp, at a record speed of 247 meters per minute.

Veracel produces bleached eucalyptus kraft pulp for export to manufacturers of tissue and printing and writing paper products in Europe, the United States, China, and Japan. About half of the mill's annual pulp production goes to Stora Enso mills in Europe and China. The company is also a major plantation holder: of its 234,000 hectares of plantation forest, about half is eucalyptus.

The Veracel facility has gained notoriety, not just because of its size but also because of allegations of corporate corruption, illegal logging, and protests over the environmental and social impacts of its mills and industrial tree plantations (for example, for displacing local people). Before the Veracel mill even opened, the Landless Workers' Movement (Brazil's largest social advocacy group) cut down 4 hectares of Veracel's plantation trees and occupied 25 hectares of the land (in April 2004). Declaring, "nobody eats eucalyptus," they planted corn and bean crops. In June 2008, a Brazilian Federal court ruled that Veracel must uproot 96,000 hectares of eucalyptus plantations and replant the land with native trees.

Despite the ongoing controversies, Stora Enso and Fibria maintain ambitious plans to double the size of the Veracel pulp facility and further expand the company's eucalyptus plantations. Loans and policy conditions, particularly through international financial organizations such as the World Bank, International Monetary Fund, and the European Investment Bank (EIB) are supporting the industrial expansion in the South. Veracel, for example, received US$30 million in loans from the EIB for plantation development and US$80 million to construct the pulp mill. As with so many Northern forest companies now working in tropical countries, the potential gains from low-cost manufacturing outweigh the potential socio-political and environmental risks.[8]

Capturing emerging markets

Multinational forest companies, then, have remained competitive by consolidating to achieve economies of scale and by strategic repositioning within lowest-cost, highest-growth markets. With growth and cost advantages shifting to the South, corporate strategies are adapting. Three new approaches stand out in particular: divesting assets in the North and reinvesting in the South; establishing supply chain linkages with China; and capitalizing on Southern export dependence.

Divesting in the North, investing in the South

Increasingly, global timber companies are divesting their older, less efficient Northern mills, and establishing departments dedicated to guiding decisions and investments in the high-growth BRIC economies of Brazil, Russia, India, and China – what they call D&E markets (developing and emerging). The major global forest companies, for example, have invested tens of millions of dollars over the last decade to expand pulp and paper production in these markets.

International Paper is illustrative. Since 2005 it has divested more than US$11.3 billion of its assets within North America (including 2.2 million hectares, comprising 85 percent of its forestlands, worth US$6 billion) while establishing new business interests in Russia, Brazil, and China. The company established a US$650 million joint venture with Ilim Holdings, Europe's largest timber producer and Russia's largest pulp, board, and paper company; it also put in place plans to invest up to US$1.5 billion in expanding paper, packaging, and market pulp production in Siberia and northern Russia. As well, it put US$140 million into a joint venture with China's Shandong Sun Paper for a coated paperboard plant to sell to core customers like Walmart and McDonald's. International Paper also recently constructed a US$1.15 billion pulp mill in Tres Lagoas, Brazil, to supply its nearby paper factory. This investment gives International Paper's three Brazilian mills control over roughly half of the office paper market in Latin America.[9]

The story is similar for other global timber companies. Since 2005, Stora Enso, for example, has poured funds into joint ventures and acquisitions in Brazil (e.g., the US$1.25 billion Veracel pulp mill project with Aracruz), Russia, and China, while divesting US$2.58 billion of its assets in North America and shutting down production capacity at its Finnish mills.

Overall, forest and paper companies are growing faster

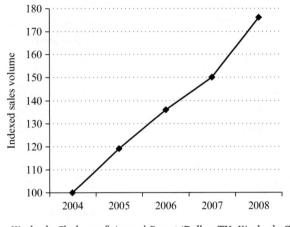

Source: Kimberly-Clark, *2008 Annual Report* (Dallas, TX: Kimberly-Clark, 2009)

Figure 3.3 Kimberly-Clark sales growth in BRICIT countries

in emerging than established markets. For example, SCA forecasts that by 2019, 80 percent of the company's growth will come from D&E markets. Brazil, Russia, India, China, Indonesia, and Turkey – what some call the BRICIT countries – will account for more than half of this D&E growth, with China providing by far the biggest sales growth of the BRICIT countries.[10] Figure 3.3 shows how Kimberly-Clark has achieved double-digit sales growth within BRICIT markets in recent years. Still, it is also of note that for Kimberly-Clark, as with other global forest companies, the majority of the company's sales (over 70 percent) remain in European and North American markets.

Linking with China
There are considerable competitive gains for companies that establish global supply chain links to China, the world's largest and fastest emerging forest and paper producer,

consumer, importer, and exporter. Optimal MNC positioning has included supplying wood fiber to meet China's large
timber deficit, capitalizing on joint-venture opportunities to
manufacture forest and paper products at low cost within the
country, and aggressive product marketing to capture a share
of the emerging Chinese market of 1.3 billion consumers.

Numerous factors are drawing MNCs into China, including low operating costs as well as efforts by the Chinese
government to remove market restrictions. The Chinese
government has also provided incentives and support for
investors to improve technology and build capacity within its
forest and paper sector. Its 2001 plan to restructure its solid
wood and pulp and paper industry was especially significant
for spurring foreign interest. The plan involved shutting
down the country's 4,000 small, non-wood (rice and straw)
pulp mills that were deemed too old, inefficient, and highly
polluting, producing small quantities of low-grade paper at
high cost. It sought foreign investment of about US$24 bil-
lion to expand wood and pulp production and establish
modern high-grade mills, with the focus on the southeastern
provinces. And it called for incentives (with the help of foreign aid) to attract foreign partners to establish 5.8 million
hectares of tree plantations to supply the expanding Chinese
mills.[11]

Since then the big Northern timber companies have been
highly active in China. They have been exporting increasing amounts of logs, pulp, wood chips, lumber, and paper to
China. Investing billions of dollars, they have relocated mills,
partnered with Chinese companies, and built new facilities.
They have invested heavily in timber plantations. And they
have branded and marketed paper, tissue, and forest products
to Chinese consumers. In the process China's forestry sector
– from pulp and paper to solid wood manufacturing – has
been transformed into a world leader.

Pulp and paper

China is now the world's second largest paper producer. Since 1990 the country has accounted for about half of the world's increase in paper production. To achieve this, China's paper industry has been shifting from agricultural residue feedstock (such as straw and reeds) to wood-based production. Before 2000, only 10 percent of China's pulp production was from wood; now it is more than 50 percent. Much of the new paper capacity in China is to produce corrugated medium and liner-board for cardboard packaging to meet the country's increased product manufacturing and export demands.

Domestic Chinese paper and packaging companies have emerged onto the global stage over the last decade (see chapter 4 for details). The Northern MNCs (including from Japan), however, have also invested billions of dollars in new, giant pulp and paper facilities in China (often integrated with timber plantations). For example, in 2002 Stora Enso established the wholly owned subsidiary, Stora Enso Guangxi, to work with a Chinese partner (Gaofeng Group) to acquire tens of thousands of hectares of land in the southern province of Guangxi to convert to eucalyptus plantation. A few years later (2006), Stora Enso acquired a site from Beihai city to establish a large integrated plantation-pulp-paper facility: one able to produce more than a million tons of pulp and close to a million tons of paper and paperboard packaging every year. Other Northern MNCs have been equally active in the Chinese market. Finland's UPM-Kymmene's Changshu paper mill in the province of Jiangsu is the biggest producer of uncoated fine paper in China. International Paper also has 11 industrial facilities to manufacture packaging across the country, producing everything from cardboard boxes to custom-designed packaging and corrugated pallets.

In addition to pulp, paper, and packaging, Procter & Gamble and Kimberly-Clark have both established factories in China to produce tissue products for the growing Chinese market. Their leading diaper brands in North America, Pampers and Huggies, are also manufactured in China. Now with the greatest brand recognition among Chinese households, these companies have managed to capture more than one third of the domestic market share.[12]

In order to meet the increasing fiber demands to expand its pulp and paper sector, the Chinese government has been supporting reforestation and plantations, even introducing a resolution for every citizen between the ages of 11 to 60 to plant 1–5 trees every March 12th on Arbour Day. China now has more than 60 million hectares of planted forest – the world's largest area of plantation. Many plantations in China, however, have been faltering – failing to meet their production targets due to poor soil conditions, inadequate silvicultural management, and high transportation costs to access remote plantation locations. Thus, the gap between domestic fiber supply and demand continues to widen, driving forest product imports (figure 3.4). China is now the world's largest commercial timber importer, purchasing wood fiber from at least 70 different countries.[13] Russia and Indonesia are China's main suppliers: a fact that introduces significant investment risks for foreign companies dealing with China, in terms of controlling for the import and processing of illegal, unsustainable timber (we will return to this topic in chapter 5 and the book's conclusion).

MNCs in China have not only had to link with logging operators and plantations in neighboring Southeast Asia and Russia, but also further afield in South America, North America, and Africa for logs, lumber, wood chips, and increasingly pulp and wastepaper imports. The top suppliers of timber to China include Russia, Malaysia, Papua New Guinea, and Gabon; Canada, Russia, and Indonesia supply close to half of China's wood pulp imports. More specifically,

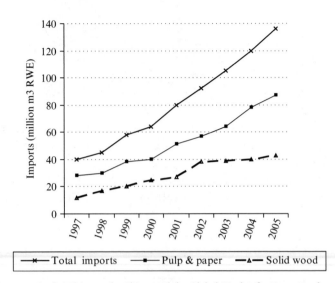

Source: Andy White et al., *China and the Global Market for Forest Products: Transforming Trade to Benefit Forests and Livelihoods* (Washington, DC: Forest Trends, March 2006; based on Chinese customs statistics), 5

Figure 3.4 China's increasing forest product imports

the company UPM-Kymmene purchases pulp from Botnia in Uruguay and then ships it to their paper mill in Changshu, China. The reason for such complex and long commodity chains is simple: it is more economical (and less risky) for UPM-Kymmene to ship pulp from fast-growing plantations thousands of kilometers away in Uruguay rather than to rely on uncertain, highly restricted fiber supplies within China.

Solid wood
The picture for solid wood in China is similar. As with pulp and paper, solid wood manufacturers depend on timber imports (particularly large-diameter raw logs) for raw material supply, and ultimately on Northern export markets for profits.

China is now, as one report sums up, the "wood workshop of the world."[14] Over the last decade the rise in production has been meteoric as large foreign investments expanded capacity. It is now a world leader in producing panels, engineered wood, furniture, moldings, and flooring; it is also a leader in secondary manufacturing of countless consumer goods, such as picture frames, chopsticks, wooden toys, pencils, and kitchen rolling pins. One result is that many of the world's major global timber supply chains now involve China-based mills and factories as Northern MNCs work to leverage the huge increase in the global capacity for low-cost wood manufacturing within China.

Historically, China's state-approved monopoly companies and a large number of small local trading houses controlled the import and export of solid wood. The government and local traders still play a significant role; yet, as China's economy has opened up, and as private companies have become more common, more opportunities now exist for direct access to overseas wood suppliers. Multinational forest companies and large global retail chains have taken particular advantage of China's greater economic openness to establish new partnerships for manufacturing solid wood. Business networks have formed among suppliers in low-cost regions (e.g., Russia and Southeast Asia), manufacturers within the highly fragmented Chinese wood sector of 200,000 small-scale operators, and buyers in the large, increasingly concentrated Northern consumer markets.

Northern MNCs have also established direct supply chain relationships within China to contract out wood processing. American forest companies now ship high-quality US logs and lumber to Chinese mills for inexpensive conversion into flooring, cabinets, and furniture products. Often, this same timber is then exported back to the United States to meet the low-cost requirements of the MNC's big retail customers.

Besides outsourcing and establishing supply chains with Chinese manufacturers, many Japanese, American, and European-owned wood product manufacturers have also relocated their plants to China – not so much to compete directly within the Chinese market, but rather to produce low-cost products for export back to home. The American-Chinese furniture industry is indicative of this trend. China is now the second largest global furniture manufacturer, with many facilities supported by US financing. More than one third of China's furniture exports go to the United States (which is the world's biggest consumer of furniture products). Similarly, in the flooring and moldings sector, Japanese-owned secondary manufacturing plants now located in China are purchasing logs and softwood lumber from Canadian forest companies, then employing inexpensive Chinese labor to convert this into flooring and door and window frames, then exporting these finished products to customers in Japan.[15]

Export dependence

Although China's consumption of forest and paper products is growing, the country's domestic industry is highly dependent on exports. Figure 3.5 shows China's skyrocketing forest and paper global exports, growing by almost 1,500 percent in dollar value since 1997. Countries like the United Kingdom now rely on China for a range of timber products, including about 40 percent of their imported hardwood plywood (by volume).

As the last chapter highlighted, such a dramatic rise in China's forest and paper exports – from well under US$1 billion to close to US$8 billion in a decade – is not surprising given the rise of big box retail and Northern consumer acceptance and growing preference for discounted products.

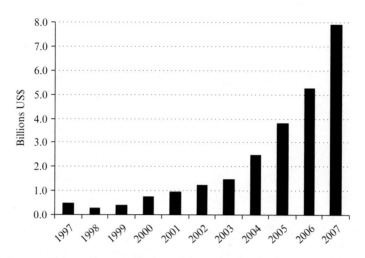

Source: Chinese Customs Data, 2008, as reported at the International Flooring Conference, Changzhou China, March 26, 2009

Figure 3.5 China's forest and paper exports

Conclusion

Chronic cycles of boom-and-bust have characterized the global timber industry over the last century. MNCs have tended to jockey for position to capture windfall profits during upswings, and have tended to restructure during and after downswings, such as the Asian financial crisis of the late 1990s and the global financial crisis of 2007–09. All have also generally followed a similar strategy to diversify and expand into overseas markets as profits and sales have narrowed in more established First World markets, and as inexpensive local timber supplies have become harder to obtain.

Today, fewer and bigger MNCs control much of the world's timber industry. The top 20, for example, account for around 60 percent of total industry sales. Timber MNCs headquartered in the global South are beginning to gain some ground, as we will show in the next chapter. But ones from the global North still dominate statistically. As of 2009, of the top 100 companies, more than three-quarters are headquartered in North America, Western Europe, and Japan. Together these Northern-based MNCs dominate worldwide timber sales as well as production. Among the top 100 forest and paper companies, for example, North American and European multinationals still account for more than 60 percent of pulp production and more than 70 percent of paper production.

Yet these First World MNCs, as this chapter shows, are increasingly relocating to the global South. As just one example, Sweden's SCA now manufactures packaging in 200 facilities across 28 countries. MNCs like SCA are also increasingly specializing and divesting, then outsourcing and contracting partners and suppliers across the global South rather than integrating vertically across business segments within specific markets. A core goal of setting up and participating in these commodity chains is to secure a low-cost fiber supply, including accessing natural forests and owning fast-growing plantations.

This trend among MNCs to relocate facilities and partner in global commodity chains with deep ties into the global South – especially in China – is driving up consumption of forest and paper products in the emerging economies: for everything from diapers to decorative table napkins to wooden housing construction. The shift to greater interregional, instead of intraregional trade, has left many Northern timber MNCs with direct control over segments of a chain, but often with less hands-on management and oversight of the overall process as Third World forest and paper companies – the subject of the next chapter – increasingly manage and supply the timber traded within these global timber chains.

CHAPTER FOUR

The Rise of the Third World

For most of the twentieth century, three-quarters of the world's industrial timber harvest was in the global North. Since the early 1990s, however, with the economic performance of developing economies outpacing developed ones, the location of harvesting, processing, trade, and consumption has been shifting from the North to the South. Third World MNCs have also become increasingly significant investors in plantations and modern mills in the global South.

The rise of China as a timber trader, processor, and investor over the last decade has had particularly striking consequences for the world timber economy: a trend of such importance for the structure of global commodity chains that we dedicate much of this chapter to exploring this. China's strategy since banning logging in its natural forests in 1998 has been to grow, not contract, its forest and paper industry. This has involved increasing timber imports, expanding domestic manufacturing, and exporting bargain-priced finished wood products to the First World. The growth in domestic processing has also allowed manufacturers and retailers in China to serve – and stimulate – rising demand within China for paper, packaging, furniture, and construction material.

Chinese companies have not achieved such quick growth alone. As we saw in the last two chapters, Northern investors and retailers have been crucial partners for sourcing fiber as well as building and operating giant processing facilities within China. Chinese firms have also been working within a network of small- and medium-sized firms across Russia, Southeast Asia, South America, and Africa to obtain a huge supply of both legal and illegal logs. With China's timber industry continuing to expand so rapidly, processors are confronting an increasingly serious shortage of wood fiber, putting even more pressure on the world's forests: especially the boreal forests of Siberia and the tropical forests of Brazil and Southeast Asia.

Chinese firms, as the two previous chapters highlighted, are increasingly core suppliers within many of the global commodity chains linking Northern MNCs and big box retailers. These chains are also connecting firms from the rest of the global South as forest and paper MNCs and retailers look for cheap timber, land, and labor. The opportunity to link into these global chains is allowing some of these smaller Third World firms to profit and expand – often at much faster rates than with the past performance of Northern MNCs. A few Third World firms are even emerging as world players. Yet, as we saw in chapter 2, linking into these global chains is also putting intense pressure on Third World firms to keep costs as low as possible through means that do not always fall within the law, including illegal fiber sourcing, tax evasion, and poor environmental, labor, and business practices.

Weak regulatory control across much of the global South compounds the risks of such business practices for forests and local communities. Other factors further heighten the challenge of governing these Third World firms, including relatively low transparency and accountability, lengthy distances from consumer markets, and murky ownership structures. The rise of Third World firms within the global timber industry, as a result, is not only causing the source of the world's timber to shift from North to South, but is also creating longer, more complex, and more obscure global commodity chains. Globalization is making forest product chains increasingly hard to govern even as more and more states, MNCs, and big box retailers are now claiming that governance is starting to strengthen.

To understand the rising importance of the Third World for the structure and governance of global commodity chains, we begin by looking in more depth at these emerging Third World corporate dynamos.

The new players

Globalization is increasingly integrating the South into consumption in the North across all global commodity chains. For timber, as we saw in the previous chapter, Northern MNCs are setting up forest operations, joint ventures, and wholly owned subsidiaries in developing countries. Northern MNCs still dominate world timber markets. But, as the number of small- and medium-sized Third World firms increases, and with a few Third World MNCs now beginning to emerge as major global players, the traditional corporate power of Northern firms is starting to diffuse across expanding trading regions and corporate networks.

Third World firms are moving every imaginable timber product – from raw logs, lumber, panels, pulp and paper, and flooring to picnic tables, wooden toys, lamp shades, lunch bags, egg cartons, garden stakes, office folders, and pencils – to their growing domestic markets as well as to the even larger Northern markets. Some of these firms are logging forest concessions or managing industrial timber plantations. Others are operating state-of-the-art mills – both publicly and privately financed – that produce high-value paper, packaging, solid wood, and finished consumer products. Firms from the BRIC economies of Brazil, Russia, India, and China play especially big roles, although some from Indonesia, Malaysia, Taiwan, Vietnam, Thailand, Chile, and Uruguay are also gaining global profile and market influence.

Table 4.1			Emerging players in the global forest and paper top 100			
Rank 2008	Rank 2005	Rank 2001	Company	Country	Estab.	Sales 2008 ($US millions)
28	39	59	Arauco	Chile	1967	3,689
35	44	52	CMPC	Chile	1920	2,945
46	69	62	Suzano	Brazil	1941	2,264
48	66	–	Shandong Chenming	China	1996	2,239
52	96	–	Nine Dragons Paper	China	1995	2,035
53	64	87	Aracruz	Brazil	1972	1,911
56	71	66	Klabin	Brazil	1899	1,725
60	77	83	Siam Pulp & Paper	Thailand	1976	1,442
62	98	90	Yuen Foong Yu Paper	Taiwan	1950	1,381
65	70	81	Votorantim (VCP)	Brazil	1998	1,366
74	–	–	Lee & Man Paper	China	1994	1,155
77	83	–	Cheng Loong	Taiwan	1959	1,087
79	87	–	Masisa SA	Chile	1920	1,054
84	–	–	Sino Forest	China	1994	901
86	–	–	Shandong Huatai Paper	China	1976	866
88	–	–	Shan Dong Sun Paper	China	1982	836
100	–	–	Ballarpur Industries	India	1945	654

Sources: PricewaterhouseCoopers (PwC), *Global Forest & Paper Survey* (2002 Edition, 2006 Edition, 2009 Edition) (Vancouver, BC: PricewaterhouseCoopers LLP), available at <www.pwc.com/gx/en/forest-paper-packaging/index.jhtml>

The size and market reach of Third World forest and paper firms have been climbing over the last decade. Just eight Third World companies were among the world's top 100 biggest forest and paper companies in 2001; 12 had made it by 2005; and 17 by 2008 (see table 4.1). While South American companies have been among the top companies for some time, the entry of China is new, with companies like Shandong Chenming and Nine Dragons Paper rocketing up the rankings after starting up in the mid-1990s.

The story of Cheung Yan, co-founder and head of Nine Dragons Paper, illustrates how quickly some Chinese timber entrepreneurs have become fabulously wealthy. The daughter of a military man and eldest of eight children, she started in 1985 by investing US$4000 to establish a recovered paper trading company in Hong Kong. Five years later, with her small business now growing well, she co-founded America Chung Nam (ACN) with her husband Ming Chung Li, turning ACN into the largest exporter of wastepaper from the United States. In 1995, she returned to China to found Nine Dragons Paper with her husband and brother, taking the company public in 2006 and building it into China's largest containerboard producer. Forbes put her net worth at US$1.7 billion in 2009; her husband and brother are also billionaires.

Going global

As we saw in the last chapter, the traditional integrated Northern multinational forest companies took over a half-century to achieve the scale necessary to expand sales and operations worldwide. In contrast, most of the emerging Third World timber companies have integrated into global commodity chains over the last two decades. In the 1990s, traditional barriers to foreign trade – including lack of financing, tariffs, and access to new technology – gave way, providing new opportunities to access world markets. This spurred many Third World forest and paper companies with a strong domestic market to focus more on global growth. This involved modernization of existing facilities, new partnerships, offshore investments, and increasing exports to First World markets.

The last decade has seen this trend continue strongly as exports from emerging economies grew at a faster rate than the global average. From 2002 to 2006, for example, exports from emerging economies grew at 15.4 percent compared to a global average of 10.6 percent. Over this time the Third World's share of timber exports went from 23.7 percent to 28.1 percent, with Third World timber and forestry companies becoming increasingly core to the functioning of global timber chains. The top Third World exporting countries in 2006 for wood, pulp, and paper were China (5.3 percent), followed by Brazil (3.9 percent), Indonesia (3.3 percent), Russia (2.5 percent), and Malaysia (2.1 percent).[1] But many other developing countries – such as Cameroon, the Congo, Chile, Gabon, Taiwan, Uruguay, and Vietnam – are playing increasingly key roles, too.

Born global

Many of the local companies exporting timber from these countries were "born global," integrating into a global commodity chain and selling overseas from the opening day of business, and not, as was the norm with most Northern MNCs, after first building up a strong domestic base.[2] By going global immediately many have been able to achieve record rates of international growth for timber companies. Asian firms, in particular from China, have been especially successful doing this. All three of China's top paper companies, for example, were born global: created from 1994 to 1996, then within a comparatively short time span, entering into the top 100 global timber multinationals with worldwide operations, partnerships, customers, and fiber sourcing (see table 4.2).

Many timber firms beyond China are being born global, too. Fibria was established in 2009 through the merger of Brazilian companies Aracruz and Votorantim Celulose e

Company	Created	Revenue (US$) & employees (2009)	Global scale
Shandong Chenming Paper	1996	$2.17 billion 16,291 employees	Leading Asian pulp and paper producer with 10 subsidiaries. China's largest papermaking enterprise and 25th biggest globally. In 2006, started up the world's largest de-inking and newsprint machine.
Nine Dragons Paper	1995	$1.98 billion 10,800 employees	China's largest containerboard producer with 99% sold within China for domestic and export markets. Also owns America Chung Nam (ACN) – the largest exporter of recovered paper from the US and a leading exporter from Europe and Asia.
Lee & Man Paper Manufacturing Ltd.	1994	$1.24 billion 7,000 employees	China's 2nd largest containerboard producer with a pulp mill in California and fiber procurement from around the world.

Table 4.2 China's born global forest and paper multinationals

Sources: Company financial reports; and *Corporate Information*, available at <www.corporateinformation.com>

Papel (VCP). From the start, the aim has been to achieve an economy of scale able to dominate in global markets. Upon its formation, Fibria became the largest "market pulp" producer in the world (*market pulp* is sold rather than used internally within an integrated pulp and paper facility) – almost double the size of its closest competitor Arauco of Chile, and triple the size of others in the global top five: APRIL of Indonesia, Sodra of Sweden, and Suzano of Brazil (figure 4.1). In its first year in operation in 2009, Fibria held a 37 percent share of the world's market pulp in eucalyptus and a 22 percent share of the world's market pulp in hardwood, controlling 11 percent

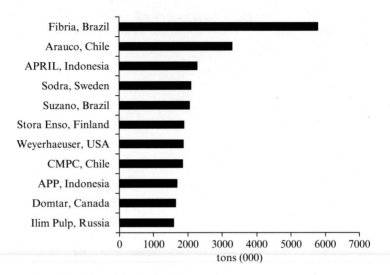

Source: Hawkins Wright, in Fibria Company Presentation, "Fibria: A Global Leader is Born," São Paulo, Brazil, September 1, 2009

Figure 4.1 Global leaders in market pulp (2009)

of total global pulp capacity.[3] Firms from developing countries are now competing strongly in this segment of the world timber industry, accounting for seven of the top 12 global market pulp producers (figure 4.1).

Born-again global

Companies born global over the last two decades are reshaping the global timber economy. But older Third World timber companies have not sat idly by ignoring the opportunities of globalizing commodity chains. Many are now what some business analysts call "born-again global," moving to "suddenly embrace rapid and dedicated internationalization" from a well established position in a domestic market.[4]

Klabin, Brazil's oldest and South America's largest paper producer, illustrates the rapid global growth trajectory of many of these born-again global firms. Mauricio Klabin founded Klabin in 1890. It started in São Paulo as a small print house and importer of office stationery. Over the next 90 years the company grew to dominate the Brazilian market through conventional strategies, such as domestic mergers and acquisitions, vertical integration, woodland purchases, and investing in plantation forests.

The company shifted suddenly toward a more global focus in the 1990s, concentrating in particular on accessing Northern markets. Global expansion activities included arranging overseas financing with the International Finance Corporation and German investors (1993–4); establishing paper production facilities in Argentina (1996); creating joint ventures with Northern integrated producers, such as Kimberly-Clark (1997), Boise-Cascade (2000), and Norske-Skog (2000); divesting lower-value business units (e.g., domestic newsprint production) to concentrate on higher-value-added paper packaging; and opening foreign sales offices to gain overseas customers. Today, according to Klabin, the company ships paper packaging material to more than 50 countries, accounting for about three-quarters of Brazil's paper-packaging exports. Klabin has also emerged as one of the world's largest manufacturers and exporters of packaging paper, accounting for 6 percent of the global market for kraft liner (strong paper used to line corrugated containers).

Many other Third World timber companies with a strong domestic presence also awoke to the opportunities of globalization in the 1990s. Born-again global Third World multinationals now among the 100 biggest forest companies in the world include: Suzano (1941) and Aracruz (1972) in Brazil; Arauco (1967), CMPC (1920), and Masisa (1920) in Chile; Yuen Foong Yu Paper (1950) and Cheng Loong (1959) in Taiwan; and Ballarpur Industries (1945) in India (see table 4.3).

Table 4.3	Major Third World forest and paper MNCs		
Company	*Created*	*Location*	*Global capacity*
Suzano Pulp and Paper	1941	Brazil	The largest Latin American paperboard producer and the second largest global eucalyptus market pulp producer (with mills in Brazil and 50% ownership of a 271,000-hectare eucalyptus plantation). Founder's son, Max Feffer, discovered how to make quality paper from eucalyptus, transforming Brazil's industry. Company had sales of US$2.16 billion in 2008, with 3,540 employees.
Arauco	1967	Chile	Pulp mills in Chile and Argentina and panel mills in Chile, Brazil, and Argentina. In 2009, jointly acquired (with Stora Enso) Ence and its plantation operations (130,000 hectares) in Uruguay, making it the largest landowner in Uruguay.
Shandong Huatai	1976	China	China's largest newsprint producer: sales were US$836 million in 2008 (with 6,484 employees). One of Asia's largest consumers of recovered fiber. Sources wastepaper from North America through its subsidiary Huatai USA.
Ballarpur Industries Ltd. (BILT)	1945	India	India's largest manufacturer and paper exporter – mainly writing and printing paper. Sales of US$605 million in 2009. In 2007, acquired Sabah Forest Industries (SFI – the largest integrated paper and pulp mill in Malaysia).
APRIL	1994	Indonesia	World's second largest producer of bleached kraft hardwood pulp. Owns plantations and manufactures pulp, paper, and paperboard mainly in Indonesia and China. Operates the world's largest pulp mill in Riau, Indonesia. Two thirds of Indonesian paper mills use APRIL pulp.

Table 4.3	(continued)		
Company	*Created*	*Location*	*Global capacity*
Asia Pulp & Paper (APP)	1994	Indonesia	Third largest global pulp, paper, and packaging capacity with assets of more than US$10 billion. Principal operations in Indonesia, but has pulp and paper investments worldwide. Owns 17 pulp and paper companies in China and 270,000 hectares of eucalyptus plantations.
Hansol-Paper	1992	Korea	A division of Korean conglomerate Hansol Corporation. Operates the largest printing and writing paper mill in Korea. Sales of US$1.2 billion in 2008 (with 980 employees). It exports half of its production.
Rimbunan Hijau (RH)	1976	Malaysia	Largest multi-industry conglomerate and integrated timber company in Malaysia. Operations in Indonesia, Papua New Guinea, Russia, New Zealand, Gabon, Equatorial Guinea, and British Guyana. Owns the Sin Chew Daily, the largest Chinese newspaper in Malaysia. Also owns RH Hypermarket Pty Ltd – the largest hypermarket retail chain in Malaysia.
Ilim	1992	Russia	One of the largest global suppliers of pulp and paper (office paper and container boxes) to the Chinese market. Largest forest resources and logging volumes in Europe and sixth largest globally. Second in Europe and seventh globally in terms of market pulp production volumes.

Going global, whether born or born-again, has allowed some Third World firms to expand operations with great force. These firms now operate some of the biggest mills in the world. The biggest "single-site" pulp complex is owned by APRIL and is located in Riau, Indonesia. It produces 2 million tonnes of kraft pulp per year. The largest "single-line" pulp mill in the world is Asia Pulp & Paper's 3,800 tonne-a-day mill on Hainan Island, China. This will soon be surpassed, however, by an even bigger single-line mill that is currently under construction in the state of Mato Grosso do Sul, Brazil, which will produce 4,380 tonnes a day (or about 1.5 million tonnes per year). The mill is owned by Eldorado Celulose e Papel Ltda (Eldorado Brasil) – a Brazilian company recently born global.

Interregional trade

The trend among Third World firms to go global is contributing to the shift in the world timber trade from intraregional to interregional. Trade in timber, as we saw in the last chapter, tripled in volume from the beginning of the 1960s until the end of the century (from about 38 million cubic meters to 114 million cubic meters). Much of the world's timber is still traded within continents. Still, across most timber supply chains – including solid wood, pulp, paper, and finished goods – the geographic reach of exchanges is extending. Between 2002 and 2006, for example, the interregional trade in wood, pulp, and paper increased by 8 percent, 5 percent, and 3 percent respectively.[5]

Complex fiber sourcing and processing arrangements that criss-cross borders and continents, especially across the global South, is now a defining feature of the world timber industry. Third World countries and firms are playing an increasingly core role as these global timber chains lengthen. The Brazilian company Fibria, for example, ships market pulp from its efficient low-cost mills in Brazil to customers in North America (30 percent), Europe (39 percent), and Asia (22 percent); only 9 percent of its production is consumed within Latin America.

The aftermath of the 2010 earthquake in Chile illustrates the interconnectedness and market dependencies across continents of the world timber economy. At the time of the earthquake, Chile accounted for nearly 9 percent of the world's market pulp production. The earthquake severely damaged many of its pulp mills, and the largest facilities were shut down. Not only were Arauco and CMPC forced to halt production, but both companies also lost inventory recently loaded onto ships destined for Asia.

As a consequence, world market pulp supplies tightened and international paper prices spiked, with cascading effects through the global economy. China's major producers of packaging papers – including Shandong Chenming, Nine Dragons Paper, and Lee & Man Paper – that depend on Chilean radiata pine softwood pulp imports (to strengthen their packaging linerboard) had to slow production and find substitute raw materials, such as recycled old corrugated container fiber. This put upward pressure on the price of old corrugated container fiber in the United States. A slowdown in linerboard production in China could even affect China's manufacturing sector if suppliers need to wait for more high-quality (e.g., strong and tear- and water-resistant) corrugated packaging to protect products like televisions, washers and dryers, and computers for the long journey to big box stores in Europe and North America. Less shipping from China could then cause further ripple effects for Northern retailers and consumers, with some products becoming unavailable and with prices rising for those products in short supply.

China, as this example shows, plays a leading role in how and why shifts now occur in the global timber economy. Low-cost manufacturers in China, well beyond the temporary struggles of producers of packaging papers, are also now facing a growing fiber deficit. Industry analysts expect China, for example, to be importing double the amount of paper-grade wood pulp by 2016 as compared to 2006. Over this same timeframe analysts are also projecting that Latin America will double total exports of paper-grade wood pulp: with China as the primary destination. Already, China is consuming more than 400 million cubic meters of timber per year, of which just half is from domestic sources. China is now the world's top importer of pulp and recycled fiber, purchasing 13.68 million tonnes of pulp in 2009 – far more than Germany was importing when it was the world's leading importer 20 years earlier (3.9 million tonnes a year in 1990).

To meet this growing Chinese demand for wood imports, many companies in the tropics, often with state and international financial support, are now rapidly clearing natural forests and exporting fast-growing plantation wood. Buyers in Latin America source fiber in particular from hardwood eucalyptus and softwood pine plantations. In Asia, however, timber manufacturing is far exceeding plantation development – a trend that is stimulating even higher and faster rates of logging in natural tropical forests.[6]

Companies in China are also now seeking new offshore fiber sources and partnerships. Shandong Sun Paper, China's largest private paper company, has plans to invest US$300–500 million in constructing pulp mills in Laos as well as another US$15 million in 100,000 hectares of eucalyptus plantations. The company has already invested US$15 million in a chipping mill in Vietnam to process the Laos wood. The plan is to export the wood chips from Vietnam to Shandong Sun Paper's plant in Yanzhou City in China, a joint venture with International Paper.[7]

Chinese companies are not only the leading importers of pulp, but are also leading the global demand and trade in recycled paper. Overall, if trends continue, within the next decade developing economies will likely account for about half of the worldwide demand for wastepaper as Third World firms continue to build on domestic market advantages to link into global commodity chains moving products to high-consuming markets.

Low-cost, emerging market advantages

Overall returns for timber investors from 2007–09 mirrored the downturn in the global economy. Within the industry, Third World MNCs have been the top performers. In 2008, the only global forest and paper companies with returns on capital greater than 10 percent were those in Latin America and China. In 2009, four out of five of the biggest merger and acquisition deals in the global forest and paper industry were in Latin America, accounting for two thirds of the overall value of deals that year. North America, meanwhile, fell from 55 percent to 5 percent of the global deal value.[8]

Some firms collapsed and many slowed production during this global downturn. The 160 year old, US-owned Pope & Talbot Company, as we noted in chapter 1, collapsed. North America's top two newsprint manufacturers – AbitibiBowater and White Birch Paper – both went under bankruptcy protection. At the same time some of the major Third World forest and paper companies demonstrated relatively strong resilience. For example, after a setback in 2008, the Chinese MNCs Nine Dragons Paper and Lee & Man Paper both rebounded in 2009 with nearly 400 percent increases in their stock values.

A combination of factors explains the resilience and continuing performance of some of the multinational timber firms in the global South. Fundamentally, beyond the complexities of currency effects and debt leveraging, lower production costs in the Third World, especially for fiber and overhead, have been critical. Greater local cultural and political understanding so as to better position their products and capture sales growth within their expanding domestic consumer markets has also provided Southern-owned companies with a key competitive advantage. These firms have also been able to retain – and even improve – their strategic positions within the global commodity chains that are supplying retail outlets in the global North: markets once served almost exclusively by big Northern companies.

Fiber, as our previous chapter documented, is the largest component of direct timber manufacturing costs. Companies working in South America and Southeast Asia have the lowest fiber costs in the world. Fast-growing plantations and vast unlogged natural forests are nearby; state taxes and fees to log and process timber tend to be very low; and state collection tends to fall well short of even these low charges. The costs in 2005 to "cut-and-haul" green wood in Brazil ranged from US$7–11 per short ton; in Sweden and Western Canada, on the other hand, it was upwards of US$25 per short ton. Together, these fiber and overhead advantages in the Third World allow manufacturers in the South to produce a much cheaper product. According to analysis by Santander bank in Brazil, at the end of 2009 Brazil's Fibria, for example, was producing pulp at US$222 per tonne: 57 percent below the average world price.[9]

Warm, moist conditions that support rich natural forests and fast-growing plantations provide further advantages for timber firms in the South. Some eucalyptus plantations in Brazil, for example, produce trees suitable for pulp production after just six to seven years and for sawn timber after just 14 years. Acacia trees planted in Indonesia can mature in only seven years: roughly seven times the growth rate of trees in Scandinavia. Softwood species, such as radiata pine in Chile, mature in 20–24 years; whereas in North America and Europe it typically takes 40–60 years.[10]

These low-cost advantages are not shared equally across all countries of the global South. Brazil, with particularly fast-growing plantations and relatively well-developed road networks, provides firms operating in Brazil with a cost advantage over other Southern states in pulp production. Indonesia, with vast natural forests and growing plantations, has gained a competitive advantage in pulp and solid wood production; China's large workforce and expanding consumer demand has provided firms working in China with an especially strong competitive advantage in paper and solid wood production and finished consumer goods. Because of its vast boreal forests and close proximity to China, Russia also has a relative low-cost advantage in pulp and solid wood production.

Overall, then, there are numerous factors that can influence the low-cost comparative and competitive advantages across developing economies. These can include the availability of fiber, technology, infrastructure, skilled labor, inexpensive energy, and political, economic, and social stability. These advantages not only differ between countries; they are also dynamic. For example, recent shifts in currency rates, energy prices, and labor costs are now encouraging manufacturers of finished wood products in China (e.g., furniture) to relocate to places like Vietnam and Cambodia.

Accelerating production

This dynamic and ongoing process of shifting competitive advantages is changing global patterns of timber supply, processing, consumption, and trade. Tropical forests, industrial plantations in the South, and the Russian boreal forest are a growing source of raw logs, pulp, and wood chips. Modern pulp, paper, and solid wood facilities in the South are operating at higher capacity and at a lower cost than aging mills in the North; consequently, Third World countries and firms are accounting for an increasing share of global timber production and exports.

This shift has been occurring over several decades. The production of commercial timber in the global South, according to the Food and Agriculture Organization, rose from 14

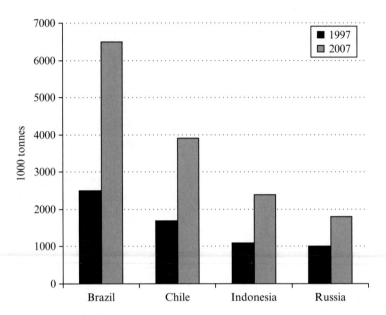

Source: Global Trade Atlas, 2009 as cited in Vincent Honnold, "Developments for the Sourcing of Raw Materials for the Production of Paper," *Journal of International Commerce and Economics,* United States International Trade Commission (August 2009): 18

Figure 4.2 Third World emerging global pulp exporting countries

percent of the world total in 1961 to 32 percent in 2008. Over this time global forest product exports from the South to the North increased significantly. Products made in the North, meanwhile, continue to be primarily traded and consumed within the North – creating an overall pattern where the North is consuming an increasing amount of the world's timber. For products such as pulp, for instance, exports from Third World countries (particularly Brazil) more than doubled from 1997 to 2007 (see figure 4.2). Regionally, over the past 50 years, Latin America's share of global timber production increased from 3 percent to 10 percent. This part of the world has been an especially valuable source of timber over the last 15 years, with the world's highest annual increase in commercial timber production from 1997 to 2005.[11]

Changes in the global paper industry illustrate these trends. Since 2000, emerging markets have been responsible for 85–90 percent of the growth in world paper markets. China has been at the forefront here. Paper production in China doubled between 2002 and 2007, while growing by only 10 percent in the rest of the world. Over this period, North America's global market share declined from 30 percent to 25 percent; China's share, meanwhile, increased from 12 percent to 19 percent.

Other BRIC countries – along with other developing countries like Indonesia, Malaysia, Taiwan, Vietnam, Cambodia, Argentina, Chile, and Uruguay – are also adding capacity for paper production, in part to replace the shutdowns in North America and Western Europe. Developing economies still account for well under half of the global paper market (38 percent). But in recent years this market has been growing steadily (by 3–6 percent), while Northern ones (including Japan) have been experiencing much slower growth rates (about 0.5 percent per year).[12] Not surprisingly, Asia (minus Japan), as figure 4.3 shows, accounts for most of the increases in papermaking capacity over the past few years.

Seeking the North

The future will more than likely see rising domestic consumption in China and India as an increasingly core driver of the global timber economy. Nonetheless, consumption in these emerging economies is still far below Northern levels and Northern MNCs still control much of the world timber market. The scale of Third World corporations (in terms of market capitalization and sales) remains much smaller than

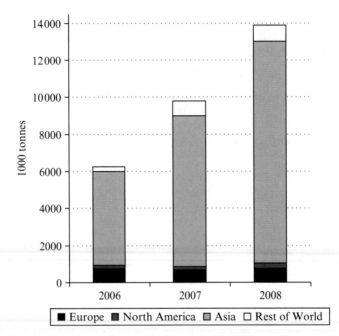

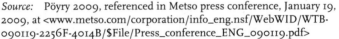

Source: Pöyry 2009, referenced in Metso press conference, January 19, 2009, at <www.metso.com/corporation/info_eng.nsf/WebWID/WTB-090119-2256F-4014B/$File/Press_conference_ENG_090119.pdf>

Figure 4.3 Global paper machine start-ups, for all grades of paper (2006–2008)

First World ones. And for most global commodity chains it is First World MNCs and big box retailers that control entry into the higher-end Northern markets. Annual sales revenue of every Third World firm, as table 4.1 showed, is still under US$4 billion; the top five North American and European MNCs, on the other hand, have annual sales ranging from US$14 to US$25 billion. Similarly, according to PricewaterhouseCoopers, while the average market capitalization in 2007 of the top American timber firms exceeded

US$90 billion and was around US$70 billion for European companies, it was still only around US$20 billion for the top companies in China and Brazil.

Third World timber firms are certainly on the rise. But, for now, even the biggest Third World MNCs still depend on access to large, higher-value Northern markets and customers. The North is also a primary source for timber manufacturing equipment and technology, exported to the South with the support of government-backed Northern export credit agencies such as the US Export-Import Bank, Euler Hermes (Germany), Exportkreditnämnden (Sweden), Finnish Guarantee Board (Finland), the Export Development Corporation (Canada), and the Export Credits Guarantee Department (United Kingdom). Southern firms seek advice from Northern timber consulting firms such as Pöyry (the world's largest forestry and engineering consulting firm, with headquarters in Finland and offices in more than 50 countries) for pulp, paper, and solid wood project assessments and design. Furthermore, Third World firms rely on Northern equipment manufacturers such as Metso (Finland) and MK Systems (US) to retrofit or construct high-capacity modern facilities.

Southern markets are consolidating in South America (e.g., Arauco and CMPC account for 80 percent of Chile's pulp production), but still remain highly fragmented in Asia where thousands of small- to medium-sized logging, processing, and manufacturing firms are operating. Asian MNCs are merging and acquiring these smaller domestic firms, but, as chapter 3 showed, many are also seeking complementary partnerships with Northern firms to gain better access to Northern markets, investment capital, and advanced technologies.

To meet Asia's growing fiber demands and mitigate the region's growing fiber deficit, Asian timber MNCs have been increasingly looking beyond neighboring countries of

Southeast Asia and Russia. In the late 1990s, Asian companies such as Rimbunan Hijau, WTK, Samling Corporation, and Fortune Timber acquired existing timber companies and forest concessions in the Brazilian Amazon. Over the last decade, Chinese and Indonesian companies have been buying up faltering North American pulp mills to secure high-quality pulp for manufacturing in China. In 2006, Asia Pulp and Paper (APP) purchased its first Canadian pulp mill in Meadow Lake Saskatchewan, which now ships all of its pulp production to China. In 2005, Lee & Man purchased the Unbl softwood mill in Somoa, California, with the aim of shipping 90 percent of production to China to produce high-quality linerboard for packaging. In 2010, Mercury Paper (a subsidiary of the Indonesian Sinar Mas Group) announced its intent to invest US$21 million in expanding its operations in the state of Virginia. Some large Chinese firms are also turning to the Northern "urban forest" for wastepaper fiber. For example, Shandong Huatai, China's largest newsprint producer, has a partnership arrangement with Huatai US (a leader in recycled paper sourcing in the United States) to facilitate transactions between American suppliers of wastepaper and the Shandong Huatai mills in China.

Asian MNCs also depend on Northern markets for a major hidden timber demand: "transit packaging." With the continuing and growing shift of global manufacturing to low-cost Asian countries, more and more of the world's packaging is coming from Chinese containerboard producers rather than from Northern integrated producers. Chinese producers need to ensure, however, that the quality of the paper packaging is high enough to meet European and North American consumer standards (in the food and beverage, electronic, and household durable industries). This is driving a rapid increase within China in the production of high-grade linerboard and corrugated medium. Demand looks set to keep rising, and industry analysts now expect China's share of world pulp to increase to more than 25 percent by 2020 (from 18 percent now). To maintain quality, Chinese firms will also need to source more pulp made from stronger fibers of tropical hardwood and Russian and Canadian boreal softwood (rather than using weaker fibers from recycled paper and non-wood sources).

Growing demand for containerboard by Asian manufacturers is allowing some Chinese firms to gain global economies of scale and higher profit margins than smaller domestic producers. For example, on the one hand, in 2007 Nine Dragons Paper, with an annual capacity of 4.5 million tonnes, achieved a profit of around RMB440 per tonne of containerboard produced; on the other hand, smaller Chinese firms with a capacity more around 0.4 million tonnes were only able to achieve profits of about RMB160 per tonne.[13] While highly profitable, large companies such as Nine Dragons Paper also have many competitors in China. This includes the Northern timber MNCs that now have packaging mills across Asia, including in China.

Joining global commodity chains, then, is creating many opportunities for Third World timber firms to grow and profit. But it also leaves them vulnerable to pressures from powerful Northern MNCs and increasingly influential retail chains – pressures that can push them to cut corners to serve the low-cost demands for the functioning of the global commodity chain.

Cutting corners in the South

Timber prices largely ignore the full environmental and social costs of production, especially in weakly governed countries without the regulatory capacity to force the internalization of these negative externalities. Third World timber suppliers need to keep costs as low as possible to win or retain customers who expect forest and paper products to be cheap (and delivered reliably in high volumes). As we saw in chapters 2 and 3, failure to do so is very risky for most suppliers as it is fairly easy for big timber companies and retailers to switch to another supplier selling an equivalent, but cheaper product.

Most of the smaller Third World firms have little power to raise prices to offset an increase in shipping, fiber, or labor costs. Instead, these firms commonly turn to cost-cutting measures. Some are able to innovate and become more efficient. But many turn to illegal and irresponsible business practices to reduce costs, including logging unsustainably, purchasing illegal timber, and failing to implement environmental safeguards or fair labor standards.[14]

Falsifying documents and audit reports to hide these poor business practices is common. So, too, is the creation of "front companies" to disguise the real source of timber products. Companies also try to pass off lower-quality hardwoods for higher-quality species, such as Asian oak and mahogany. Such practices create significant challenges for the transparency and accountability of Third World timber firms. Complex organizational and ownership structures, along with weak governance capacity within many developing states, further hinder transparency and accountability. Third World timber firms commonly involve intricate webs of overlapping subsidiaries and shareholders. Many are privately owned and financed and therefore do not publicly disclose their finances or holdings. Company records are often outdated, too. Moreover, it is not uncommon for these companies to have undisclosed buyouts, use tax havens and proxy directors, and employ a complicated strategy of company listings, de-listings, re-listings, and secondary offerings on various stock exchanges. All of these practices obscure some basic company information: such as ownership, profits, and sourcing locations.

For example, the major global companies Asia Pulp and Paper (Singapore), APRIL (Singapore), and Ilim (Russia) were all listed in PricewaterhouseCooper's 2006 global top 100 timber companies. Yet in 2007 these firms suddenly dropped off: de-listed from their respective exchanges and no longer disclosing any financial information. On the surface this may make it appear that Russia does not have any major global forest companies. Yet Ilim is Europe's largest solid wood producer and second largest pulp producer, with annual sales of just under US$2 billion a year and a lease of over 5.3 million hectares of the Russian boreal forest. It simply disappeared from the global ranking when it went private. The companies APRIL and Asia Pulp and Paper are also private, but remain the parent companies for many listed subsidiaries along the forestry supply chain. Asia Pulp and Paper is the parent, for example, of Indah Kiat and Tjiwi Kimia: two large Indonesian logging companies listed on the Jakarta and Surabaya stock exchanges.

A closer look at Asia Pulp and Paper further illustrates the organizational complexity of many Third World companies. Singapore's Asia Pulp and Paper is linked to Indonesia's Sinar Mas Group and represents the brand of paper products produced by the APP Group, including APP China. The company is structured as a brand umbrella for a large number of independently owned and operated mills around the world. This includes Mercury Paper in the United States and Paper Excellence B.V. in Canada. And it also includes many Southeast Asian companies operating on the Indonesian islands of Java and Sumatra. Asia Pulp and Paper's timber supply chain is also complicated and difficult to trace, as the company does not own any mills or forestry operations, but rather relies on numerous "exclusive pulpwood" suppliers.[15]

Conclusion

Third World MNCs, then, are no doubt gaining ground in the global timber industry. The reasons vary across companies and countries. But core factors include easier access to low-cost fiber, labor, and infrastructure, more opportunities to expand in rapidly growing domestic markets, and similar, if not better, access than traditional Northern MNCs to high-end consumer markets through globalizing commodity chains. Linking into these global chains has allowed many to grow over the last decade at a much faster rate than Northern MNCs did over the last century. Also, cost-saving ability at home has allowed some of the emerging Third World MNCs, such as China's Nine Dragons Paper, to fare much better than Northern MNCs during economic downturns, such as the 2007–09 global financial crisis.

Some of the Third World MNCs now emerging from Asia, South America, and Russia are also starting to look different from the smaller timber firms working in the global South. These bigger companies are more vertically integrated in terms of producing higher-value timber products for domestic consumption and export. Many are also investing in huge, modern sawmills, plywood plants, paper machines, and packaging plants around the world, including in the North. They are collaborating with larger established multinationals. For example, Lee & Man Paper has established strategic partnerships with Yuen Foong Yu Paper (Taiwan's largest paper manufacturer) and Japan's Nippon Paper. Some, too, are growing by consolidating with other Southern firms. As we saw in this chapter, for example, two of Brazil's largest pulp producers – Votorantim Cululose e Papel and Aracruz Cellulose – merged in 2009 to create Fibria: in the process becoming the world's largest and lowest-cost market pulp producer.

Some Third World MNCs are also investing heavily in the branding and marketing of their own forest and paper products to capture the expanding per capita rates of consumption within emerging economies. As reported in the *Taiwan Economic News*, the company Asia Pulp and Paper, for instance, budgeted more than US$6 million in 2010 to advertise and promote its household paper in rapidly growing markets such as Taiwan. In fact, Asia Pulp and Paper created new brand names just for Taiwan (e.g., Paseo, VirJoy, and LiVi), with the goal of capturing 15–20 percent of the Taiwanese market. Companies see great market growth potential here: the average consumption of household paper in Taiwan is around 7.7 kg per person per year, less than half of a typical European, American, or Japanese consumer. Asia Pulp and Paper's multimillion dollar advertising budget is unprecedented among the traditional paper companies in Taiwan (e.g., Yeong Foong Yu and Cheng Loong) and many industry analysts are now expecting significant shifts in Taiwan's household paper markets, including rising rates of consumption.

Still, even as more Third World MNCs move into the 100 top global timber companies, all of them remain much smaller than the top 10 Northern ones – and the vast majority continue to be weak players in high-end Northern markets. Suppliers from the Third World are at a structural disadvantage and often rely on the financing and markets of Northern MNCs and big box retailers. Chinese paper companies, for example, have been modernizing and adapting mills over the last decade to supply higher-grade linerboard and corrugated paper to meet the specifications of Northern buyers, particularly the large retail chains. To produce higher-grade containerboard, the Chinese mills also need higher-grade pulp, requiring them to switch from using non-wood materials, recycled paper, and plantation timber (with shorter, weaker fibers) to higher-grade market pulp (e.g., longer, stronger wood fibers) to blend into their paper mix. This includes sourcing a higher percentage of natural tropical hardwood pulp as well as mainly softwood pulp from the native forests of Canada and Russia.

After investing in the upgrades and fiber to link into a global commodity chain, most Third World companies cannot afford to even risk a buyer switching them for another supplier. Power within most global commodity chains therefore continues to rest with the Northern MNCs and big box retail buyers – and, even as many Third World companies gain market strength and earn reasonable profits, only the rare few are in the global top 50 and able to wield much influence over chain structures or corporate discourse.

Instead, most Third World firms – especially the tens of thousands of small- and medium-sized loggers, processors, and traders – are under constant pressure to lower costs to maintain a toehold within these global commodity chains. As we will show in the next chapter, the low-priced goods come at a high expense. And these costs arise not only from consuming products made from this timber – a desk, diaper, or diary – but equally from products that rely on the deforested land – a leather purse, a tub of margarine, or a block of tofu.

CHAPTER FIVE

Consuming the South

Wood formed the foundation of nearly every society from the early Bronze Age of 3500 BC until the middle of the nineteenth century. Controlling timber supplies allowed societies to grow; declining supplies caused others to decay. "Wood," writes historian John Perlin, "is the unsung hero of the technological revolution that has brought us from a stone and bone culture to our present age."[1] Without it, some early societies could not even survive, as the fighting and cannibalism confirmed for Dutch explorer Jacob Roggeveen when he landed on a deforested Easter Island in 1722.

Timber continues to be a vital resource for societal well-being, environmental sustainability, and economic prosperity. The remaining natural forests that supply most of the world's timber also house much of the earth's biodiversity. They provide critical ecological services, from stabilizing soil to purifying water to regulating the climate. And they remain a necessity for many of the world's poorest people – essential for cooking, heating, and shelter.

As in the past, greed and power continue to motivate many who trade, steal, and stockpile timber. But, as chapters 2–4 revealed, the context for the struggles for control is now much different. Every forest, no matter how remote, is part of a globalized world economy. Timber flows through complex commodity chains; volumes are far higher; and international markets largely determine the value of the fiber. It is a truly global and commercial resource, traded and turned into thousands of different consumer products. For local people, however, the effects of overexploiting a forest remain much the same as in the past, especially in the global South: some short-term income and infrastructure gains, but in the longer term ruined environments, declining living standards, and social upheaval. Yet, as this chapter documents, the consequences of today's global political economy of timber for the nature of consumption and the state of the global environment are now on a scale and intensity far greater than anyone could even have imagined possible a few hundred years ago.

The picture that emerges is stark: corporations are mining a lucrative commodity in the poorest and most vulnerable regions of the South to fuel emerging economies, sustain rising demand for timber and products from deforested tropical land, and serve bargain-priced and wasteful consumption in the North.

Rising consumption of timber

From the early 1960s to 2007, global timber consumption increased by about 70 percent – from about 1 billion cubic meters to 1.7 billion cubic meters per year. The scale of current timber consumption is extraordinary: 1.7 billion cubic meters of timber is enough to build a 40 meter wide by 1 meter high boardwalk around the earth's equator. Over the past 50 years, the uses and manufacturing of timber have expanded as well as shifted. Since 2000, less timber has been going to newsprint; at the same time demand for writing paper, tissue products, wood-based panels (e.g., plywood and particleboard), and paper packaging has been rapidly rising. The overall increase in timber consumption has tracked the growing world population and rising incomes. Put simply, as communities grow bigger and richer, people construct more homes and buildings, buy more furniture, and use more paper and packaging.

While demand has increased, over the last half-century several factors have also moderated timber consumption somewhat. Advances in technology and management have allowed firms to produce more products per tree. The recycling of paper and paperboard has been increasing. And electronic media and substitute construction materials (such as steel, concrete, and plastic) have shifted demand. Nonetheless, most analysts still expect timber consumption to keep rising. By 2030, the Food and Agriculture Organization expects industrial roundwood consumption to double (from a 2005 baseline). The reason is simple: fundamentally, timber remains an "essential" resource. The American Forest & Paper Association explains:

> There isn't a day or a minute that goes by when a forest product isn't part of our lives. The newspaper in the morning. The table where we eat our breakfast and the box that holds the cereal. The desks we work at and the paper in the copying machine. Our children's school books. The beds we sleep in and the houses that shelter us.[2]

Consumption of this essential resource is highly unequal, however, with the North consuming about 70 percent of the world's commercial timber. Per capita figures for paper and paperboard are revealing. Global paper and paperboard production, as figures 5.1 and 5.2 show, increased by more than 230 percent from 1967 to 2007. The US and Western Europe have been driving forces behind this rising paper usage. People in industrialized countries consume on average more than 200 kilograms of paper products per year while the average in developing countries has been just 17 kilograms.

This does not mean that people in the developing world do not rely on forests and wood. Most, in fact, depend far more on forests than those living in the North. The majority of people

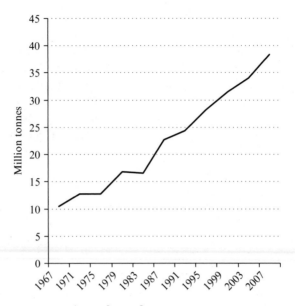

Source: FAOstat, at <http://faostat.fao.org>

Figure 5.1 Global paper and paperboard production

in the developing world, however, cannot access or afford commercial timber, especially high-end products. Far fewer would see forests as an essential resource for the "newspaper in the morning" or "the box that holds the cereal." Instead, most would see forests as an essential source of wood to heat homes and cook meals – gathering and consuming it before it ever goes to "market." In parts of Africa, for example, as much as 90 percent of the total wood harvested is still used as fuel for subsistence living.

As well, as incomes rise, per capita consumption of commercial timber is also increasing in the global South. And the wide gap in South–North per capita rates suggests a potential for a rapid increase in future consumption – especially within

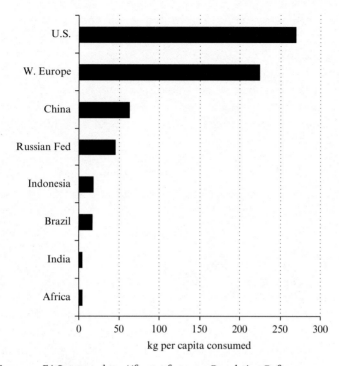

Sources: FAOstat, at <http://faostat.fao.org>; Population Reference Bureau, at <www.prb.org>

Figure 5.2 Global per capita paper and paperboard consumption, 2008

emerging economies such as China and India. Already, with a population of 1.3 billion, China's total paper and paperboard consumption exceeds that of the United States (although, as already noted, much is re-exported to the North in the form of packaging). Migrating, growing, and prospering populations in the South will need to build 1 billion or so new homes over the next 50 years. In addition, newly introduced paper products like diapers and 2-ply toilet paper are rapidly gaining popular acceptance.

For now, however, the global North still consumes most of the world's commercial timber. Increasingly, as we document next, much of this is wasteful: high volumes of printed retail catalogues and print advertising, throw-away office paper and paper packaging, disposable shipping pallets, and cheap self-assembly furniture. Furthermore, as the following section documents, the source of wood fiber for many of these products is shifting from highly managed Northern forests to the more vulnerable, under-regulated, and underpriced tropical forests of Southeast Asia, Africa, and the South American Amazon and the encroaching industrial plantations – driving up the global ecological and social costs of the world's timber economy.

Print material

Just 10 percent of the world's population living in Western Europe and North America consumes more than half of global paper production. Of this, 30–40 percent ends up in a landfill. Recycling is certainly increasing – and efforts, such as Walmart's packaging scorecard alongside government initiatives such as the EU packaging directive, are helping here. Yet paper and paper products – including newspapers, magazines, milk cartons, cardboard boxes, and office paper – remain the largest contributor to municipal solid waste.[3]

Americans are the biggest global per capita paper consumers, using more than 80 million tonnes a year. For office paper alone, Americans consume enough to construct annually "a ten-foot-high wall from New York City all the way to Tokyo." Retail catalogues and advertisements add to this, clogging mailboxes – and before long the landfills. According to the US Postal Service, American households receive more than 100 billion pieces of "junk mail" advertising a year – much of which is tossed away without ever being opened. One survey by the Consumer Research Institute in New York State found that 44 percent of junk mail was never opened. Such waste comes with a cost beyond just garbage: producing throw-away junk mail just for the American market is using up 100 million trees a year.[4]

Packaging

The biggest share of paper consumption and waste, however, is for packaging, and not for print media. Packaging is important for product quality, safety, and marketing. It protects goods in transit, storage, and distribution. And it conveys information about the product to the consumer, providing a potential competitive advantage at the point of sale by differentiating products with color and design. Paper and paperboard packaging, according to the World Packaging Organization, constitutes the largest share (40 percent) of the US$475 billion global packaging market (2003 figures): ahead of options like plastic, metal, glass, and solid wood. In the US alone, producers manufacture about 250 million cardboard boxes *every day*. While paper packaging provides benefits, it is also the fastest-growing segment of the municipal waste stream.

Solid wood packaging – especially wooden crates and pallets – is also a growing contributor to wasteful timber consumption. "Pallets," write Virginia Tech wood packaging experts Marshall White and Peter Hamner, "move the world."[5] At some point before reaching consumers most products – food, beverages, cosmetics, appliances, and electronics – are packaged in corrugated containers, placed on square, flat pallets, and covered in plastic stretch wrap. This allows forklifts to load many different products quickly and easily. Pallets can be made from plastic and steel. Still, according to the US National Wooden Pallet & Container Association, more than 90 percent are made from wood.

Every year billions of pallets are made globally. The US alone makes more than 700 million pallets a year. Although manufacturers use lower grades of wood and trimmings, at 15 board feet per pallet (0.00235 m3) this is roughly equal to one third of the hardwood lumber produced in the United States – making it the single largest hardwood lumber use in the country (most pallets in the US are made from hardwood rather than softwood for strength properties).

Why is the consumption of pallets so high? One reason is the growing global trade and movement of goods. Another is the very short lifespan of an average export pallet from manufacture to landfill – with most wooden pallets used just once. Despite some advances in recycling efforts in recent years, large quantities of wooden pallets continue to be dumped into municipal landfills every year, accounting, for example, for about 2–3 percent of the total solid waste stream in the United States.[6]

Flat-pack furniture
IKEA introduced the concept of ready-to-assemble discount furniture in the early 1960s. The founder, Ingvar Kamprad, also made himself and his family extraordinarily wealthy: sitting in 11th place on the 2010 Forbes list of the world's billionaires. In Kamprad's 1976 manifesto, "A Furniture Dealer's Testament," he claims that democratization was the real concept behind his retail strategy: "to create a better everyday life for the many . . . by offering a wide range of well-designed, functional home furnishing products at prices so low that as many people as possible will be able to afford them."[7]

IKEA's rapid growth and expansion beyond Sweden showed other manufacturers the competitive gains and savings possible from shipping, warehousing, and manufacturing self-assembly furniture. The "flat-pack" furniture market subsequently took off and market growth continues as the automation of manufacturing plants – including advanced robotics – has enabled even faster, lower-cost, high-volume production. The low production costs allow for consumer prices that are so low that no traditional furniture maker can compete on price terms.

To some extent self-assembly furniture saves forest resources. Much of the furniture is made from glued, engineered wood that uses fewer trees than solid wood pieces. Instead of using more expensive wood veneer, IKEA even lacquers photocopied images of wood fiber onto the surface of particleboard tables, dressers, and bookcases. Yet, although flat-pack furniture can use less wood, it is also less durable – prone to chipping, warping, and losing legs. Such furniture is rarely handed down and, unlike crafted furniture, it does not increase in value through generations: generally, owners assemble it once and throw it away when moving as shipping costs can be greater than the cost of buying a new piece. Moreover, flat-pack furniture production is driving up the production and consumption of corrugated packaging to transport the individually bound units. In the case of China, for example, as manufacturing of flat-pack furniture steadily grew so has China's containerboard production: from less than 5 million tonnes in 1995 to more than 30 million tonnes today.

And, finally, the high-volume, competitive economies of scale of the flat-pack industry has shifted furniture production from traditional craftsman regions like Italy and Germany to low-cost countries such as China, Vietnam, Indonesia, Malaysia, and Thailand: all, as we have seen, with some of the world's highest incidence of illegal and unsustainably sourced timber. China is now the world's leading producer of furniture, surpassing Canada in 2000 as the top wood furniture exporter to the world's biggest market, the United States – a feat it was able to accomplish by selling wood furniture at prices one quarter to one third below traditional producers. China now accounts for more than half of US furniture imports.

So, while just a few decades ago most furniture was made to order (locally), with consumers carefully choosing the wood, color, and texture, now the furniture industry is global, dominated by mass manufacturing (especially in Asia), bulk orders from big box retailers, and bargain prices for consumers in the North. Most of these consumers are blind to the origin of the wood or its possible connection to the loss of tropical or boreal forests. And this is true not only for furniture. Low-cost timber exports from developing countries are increasing across all sectors of the industry. China is now among the top global players in panel, sawnwood, and secondary manufactured wood product markets, such as flooring, decking, and moldings, as well as an emerging leader in paper export markets.

Rising tropical timber sourcing and trade

Over the last few decades, as we saw in chapters 3 and 4, the global timber industry has been gradually shifting from the temperate and boreal forests of North America and Europe to tropical forests and plantations in the South. Seeking greater returns, investors have been migrating from well-established timber economies (now with relatively little undisturbed forest) to the poorer regions of Asia, Africa, and Latin America with large, untapped old-growth tropical forests and lower-cost, faster-growing plantations. Russia's vast boreal forests are also once again becoming an increasingly core source of the world's timber supply.

Much of this timber is flowing into China. Two thirds of Russia's log exports go to China. And, as table 5.1 shows, millions of cubic meters of tropical logs now flow into China from countries such as Papua New Guinea, the Solomon Islands, Malaysia, and Gabon. (India is also a major destination for logs from Malaysia.) Table 5.1 does not factor in large,

| Table 5.1   The top global trade flows in tropical logs, 2007 ||
Trade flow	Volume m3
Papua New Guinea ➤ China	2,300,000
Malaysia ➤ India	1,500,000
Malaysia ➤ China	1,300,000
Gabon ➤ China	1,100,000
Solomon Islands ➤ China	1,000,000

Source: ITTO, *Annual Review and Assessment of the World Timber Situation* (Yokohama, JPN: International Tropical Timber Organization, 2008), 12

unrecorded illegal trade flows of tropical logs – much of which also lands in China. No one knows for sure the exact amounts of illegal timber: in some ways this trade resembles the global trade in illicit drugs. Roughly half of the world's traded timber is flowing through China. Even based on China customs data, it would seem that about 30–40 percent of this raw material supply is likely from illegal origins and 40 percent of China's timber exports comprise illegal timber.[8]

China is still behind the United States as a consumer of timber, but is now ahead of Germany, Japan, and the United Kingdom. India is also emerging as a major timber consumer, and as of 2007 is already the second largest importer of raw tropical logs, after China, and ahead of Japan and Taiwan. Such national statistics on timber consumption, however, need further unpacking. As we mentioned earlier, "consumption" of timber in China, for example, conceals significant re-exports. A huge amount of paper and wood are exported as packaging and pallets with consumer goods; and as exports of manufactured retail goods increase so does the amount of "wood" for items like picture frames, tools, toys, and wooden clothes hangers.

Trade in timber is shifting, too: away from traditional

Northern exporters, such as Canada and Finland, and toward emerging economies, such as Brazil, China, Indonesia, and Russia. As tracked by the Food and Agriculture Organization, it is shifting as well toward paper and paperboard, which by value now comprises around half of world trade in forest products. The proportion of trade in lower-value raw logs is also declining as trade in more processed higher-value timber products rises, such as furniture and flooring from China, Indonesia, and Vietnam, and pulp and paper products from Brazil. For exporters of tropical timber, raw logs now constitute only about one quarter of the total (both by value and roundwood equivalent), down from over 60 percent in the 1980s.[9]

Bans and tariffs on raw log exports account in part for this shift toward more higher-end trade. Illegal trade in logs and rough lumber has, however, become a significant portion of world trade. The Organisation for Economic Co-operation and Development (OECD) estimates that 5–10 percent of global timber production is now illegal. Much of this timber is flowing from Russia, Africa, and Southeast Asia into China where it is turned into "legal" products for worldwide export. This increasingly includes semi-processed timber commodities such as plywood and printing paper, as well as higher-end consumer products such as self-assembly furniture.

Determining for certain how much illegal timber makes its way to consumers in places like Europe and North America is impossible given the range and complexity of products and recording inaccuracies at just about every point in the long and complex global commodity chains. WWF analysts have tried to get a rough idea for Europe, estimating that as much as 28 percent of wood imports into the European Union "derive from illegal or suspicious sources," with most of this from Russia and Southeast Asia.[10] Suspicious products flowing into the EU include Brazilian walnut flooring, meranti doors and window frames from Indonesia, teak wood from Burma, and outside decking made from Malaysian merbau. The United States, according to the International Tropical Timber Organization, imported about US$3.6 billion of hardwood products in 2006. Of this, 30 percent was of tropical origin from countries such as China, Brazil, Indonesia, Malaysia, Taiwan, Ghana, and Cameroon. Studies on the illegal timber trade by organizations such as the OECD (in 2007) estimate that about 10 percent of US imports of tropical timber products are of questionable legal origin.

Governments have recently introduced initiatives such as amendments to the US Lacey Act, the European Parliament's "due diligence" legislation, and the Forest Law Enforcement, Governance and Trade (FLEGT) bilateral voluntary partnership programs to help curtail the flood of illegal timber imports that are now ending up in North American and European markets, and particularly on the shelves of big box retail stores. As chapters 2–4 showed, by shifting from domestic manufacturers to lower-cost overseas suppliers in countries like China, the big retail chains have been able to decrease prices. But this strategy also introduces a significant risk of selling illegal and unsustainably produced timber. As the next section documents, the world's biggest supermarket and hypermarket retail stores are increasingly facing similar risks, not just of stocking shelves with unsustainable timber products, but also unsustainable soy, beef, and palm oil grocery products sourced from recently deforested tropical rainforests.

Hidden costs of groceries

Global commodity booms in food and fuel are adding to the pressure on the forests of the global South. Responding to market signals over the last decade, plantation owners in the

South have been clearing (and burning) tropical forests to make way for cattle ranching and industrial crops. Land use decisions in the South are now closely linked to fluctuating global prices, global commodity chains, and shifting consumptive demands of Northern markets.

The most marked boom of the past century (in terms of a sharp and sustained rise in global prices) was from early 2003 to mid-2008. Over this time, world energy prices increased by 320 percent, metals and minerals by 296 percent, and globally traded food prices by 138 percent.[11] Soaring world prices for agricultural commodities spurred further expansion of mechanized, large-scale oil palm plantations in Indonesia and Malaysia. This also accelerated the expansion of soybean crops and cattle ranching in Brazil – destroying vast tracts of tropical forest in the process. As a result, global production of palm oil and soy has been rising quickly over the last decade, with shelf after shelf of the big box stores in the North now containing these as ingredients.

Production

Global production of palm oil has reached record levels in recent years alongside high rates of tropical forest loss (figure 5.3). Indonesia and Malaysia are the world's biggest palm oil producers, together accounting for 85 percent of world production and 88 percent of global exports. Over the past decade the area covered by oil palm in Indonesia quadrupled to more than 7 million hectares. In Malaysia, meanwhile, it has expanded threefold since 1980, now occupying 4.5 million hectares (more than 13 percent of the country's total land area). Around half of the palm oil plantations in these countries have relied on clearing tropical forests. Some palm oil and timber companies even overlap. The Sinar Mas conglomerate, for example, is the largest palm oil producer in Indonesia and the second largest in the world. Through its

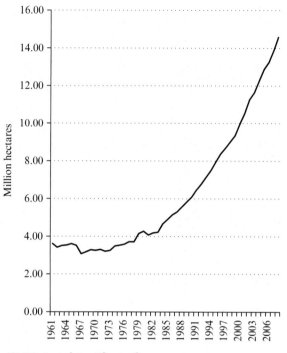

Source: FAOStat, at <http://faostat.fao.org>

Figure 5.3 World oil palm plantation (area harvested)

subsidiary Asia Pulp and Paper, it is also one of the world's biggest integrated pulp and paper companies, with significant holdings of natural forestland in Indonesia.

Similar to oil palm, soybean production has grown rapidly since 2000 in response to high global demand (figure 5.4). Just behind the US, Brazil has emerged as the world's second largest soybean producer (Argentina, China, and India are also top global producers). Brazil now accounts for just over one quarter of global production. And its market share looks set to keep rising as American production drops due

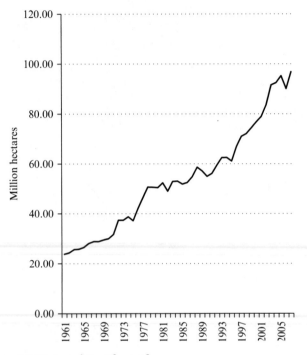

Source: FAOStat, at <http://faostat.fao.org>

Figure 5.4 World soybean production (area harvested)

to farmers shifting from soybeans to corn for ethanol. In 1940, Brazil had only about 700 hectares of soy fields. Led by the American multinational companies ADM, Bunge, and Cargill, the country has planted about 22 million hectares with soybean crops – an area the size of Tennessee and Kentucky combined.

Much of this is on land that not long ago was Amazon rainforest. For instance, 20 years ago the state of Mato Grosso in the Amazon region had no soy plantations: today, it is Brazil's largest soy-producing state. Soy is currently the most significant plantation crop driving deforestation of the Brazilian Amazon. But oil palm also plays a role. The Brazilian government recently announced, for example, a plan to provide US$60 million in subsidies to encourage the development of oil palm plantations on 50 million hectares of "degraded" Amazon forestland.[12]

Clearing for cattle ranching is causing even more deforestation in the Brazilian Amazon than clearing for soybean and oil palm plantations. There are now more than 10,000 cattle ranches in the Brazilian Amazon alone. These cover 8–9 million hectares and directly account for more than half of the country's forest loss since the mid-1990s – making cattle ranching in the Amazon one of the world's biggest drivers of deforestation. Brazil now has the world's largest commercial cattle herd with at least 87 million cattle grazing in the Amazon (up from 57 million in 2002) and more than 180 million in total. It is also the world's largest beef and tan leather exporter (by volume). JBS, the world's largest corporation in the global protein industry and the world's biggest beef producer, is processing more than 43,000 heads of cattle every day in Brazil (as of January 2010). JBS is also the world's largest leather producer, supplying nearly 150,000 square meters of Brazilian cattle hides per day. The fastest growth in Brazil's cattle herd is occurring in the Amazon states of Mato Grosso, Para, and Rondônia – the Brazilian states with the most deforestation, forest fires, and illegal land use.

Similar to soy and oil palm plantations, overseas consumption has been a major driver of Amazon forest conversion for beef and leather. By 2004, beef exports had risen fivefold (from 1997 levels) and the country had surpassed Australia as the top global beef exporter by volume (figure 5.6). Around 40 percent of this beef from 1999 to 2002 went to countries in the European Union. Meanwhile, Amazon deforestation was increasing – by then well over 2 million hectares per year.

Consumption

Tropical forest leather, beef, palm, and soy end up in many products. Bovine (cow) leather is a high-end textile for clothing, upholstery, and auto interiors. Palm oil is widespread in processed foods, such as chips, crackers, soups, cooking oil, margarine, and chocolate, as well as cosmetics, soaps, industrial lubricants, and biofuel. Soy products include tofu, soy sauce, and margarine; it is also a common filler in many processed foods. Soy is produced mainly, however, for animal feed (80 percent of global production) and vegetable oil. Soy and palm oil are now the most consumed vegetable oils in the world – holding about one quarter and one third of the market share respectively.

Global demand for soybean and palm oil continues to increase. Livestock around the world is increasingly fed with soymeal. Soy is also the single most common ingredient in foods processed in the US and is found in 60 percent of processed foods in Europe. Palm oil, meanwhile, is in 10 percent of grocery products in places like Europe and North America, including in many margarines. Margarine is now a major part of the Western diet, with the average American, for example, consuming more than 3.5 kilograms per year (with the country importing about 4.5 billion kilograms a year). Food industry consumption of palm oil is rising. Demand is also starting to expand in biofuel markets, with some environmentalists now referring to palm oil as "deforestation diesel."[13]

The company Unilever is currently the world's largest consumer of palm oil. Although its global purchases continue to increase, in 2009 the company did announce that it would stop buying palm oil from Indonesia's largest palm oil producer, Sinar Mas, following allegations that the company was illegally clearing rainforests in West Kalimantan to plant oil palm. (Specifically, Greenpeace accused Sinar Mas of logging without permits, damaging high-biodiversity areas, and illegally draining peatlands.) Nestlé also recently followed Unilever's lead, dropping from their supply chain oil palm and paper companies like Sinar Mas and others with alleged links to deforestation.

Global meat consumption has increased almost sixfold over the last 50 years. And it is on track to double again by the middle of this century. Already, there are more than 1 billion pigs, 1.3 billion cattle, 1.8 billion sheep and goats, and 15 billion chickens. Beef accounts for about one quarter of the global meat consumption (by weight). The US is the world's largest per capita consumer of meat, averaging over 125 kilograms, or 275 pounds, a year. Beef is at the heart of this meat diet and the US is the world's largest beef consumer; however, growth rates are highest in developing countries with historically low per capita rates of consumption, particularly China.

Markets for tanned leather are also expanding. According to the International Council of Tanners, roughly two thirds of the global leather market is bovine leather and just over half of all leather goes into footwear. China is the largest shoe manufacturer with about 60 percent of the global market. It imports tanned leather from companies like Bertin in Brazil to produce top brand-name running shoes for export, especially to North America and Europe. Brazil also exports leather for clothing, furniture, and automobile upholstery, with exports to Italy, the US, and China, for example, generating about one quarter of its cattle industry profits.

Export markets are fundamental drivers of the rising production of beef, leather, soy, and palm oil on the deforested land of the global South. Indonesia exports about 60 percent of its palm oil production; figures 5.5 and 5.6 illustrate the rise in Brazilian soy and beef exports. European consumers play a key role here. The EU is now consuming about one third of Brazil's beef, one third of the world's soy, and 15–20 percent

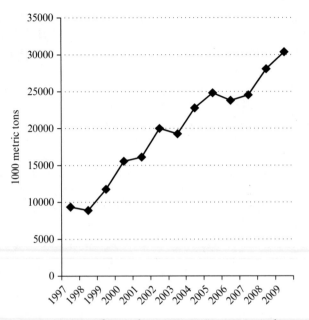

Source: US Department of Agriculture (USDA) Foreign Agriculture
Service (FAS) database, at <www.fas.usda.gov>

Figure 5.5 Brazil soybean exports (1997–2009)

of global palm oil. China, as with timber, is also at the core of
this trade, now the world's biggest importer of soy and largest
consumer of palm oil.

In short, grocery, fast food, clothing, furniture, and auto
sectors are all major buyers and sellers of products grown
on deforested tropical lands. The list of just leather items is
long: from luxury leather seats, recliners, and sofas to cheap
footwear, handbags, and gloves. Fast-food chains are supplied with millions of tonnes of chicken and beef fed with
soy from deforested land. And the grocery aisles of multinational supermarket and hypermarket chains such as Walmart,
Tesco, Costco, and Carrefour are stacked with thousands of

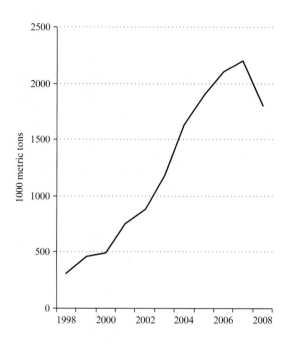

Source: US Foreign Agriculture Service, *Livestock and Poultry: World Markets and Trade* (Washington, DC: USDA FAS, October 2009)

Figure 5.6 Brazil beef exports (1998–2008)

products with "forest risk commodities," including soy, palm oil, Amazon beef, and timber. Such blind consumption is, as we show next, consuming the forests of the global South.

Deforesting the Third World

Over the last half-century *total* global forest cover has been fairly stable. Hidden in this optimistic statistic are critical changes to forest integrity and timber supplies. Secondary and plantation forests are slowly replacing primary ones.

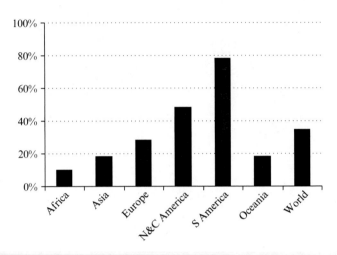

Source: FAO, *Global Forest Resources Assessment 2005: Progress Towards Sustainable Forest Management* (Rome: FAO, 2006), 41

Figure 5.7 Primary forest area by region

More than half of the world's forests (60 percent) are now secondary forests; 35 percent are primary (figure 5.7); and 5 percent are plantation forests. The location of the forest cover is shifting, too. Through reforestation, afforestation, and sil-vicultural management, Europe's forests are growing and North America's are holding steady. At the same time the for-ests of Africa, Asia, Oceania, and South and Central America are decreasing, with especially great losses in two of the world's greatest biodiversity locations: Brazil and Indonesia (see figure 5.8).

Commercial logging is certainly causing deterioration of the flora, fauna, and forest structure in the remaining ancient temperate rainforests and old-growth Canadian and Russian boreal forests (although it may improve in Canada with the signing in 2010 of the Canadian Boreal Forest Agreement to protect 72 million hectares of northern forest). Still, two thirds

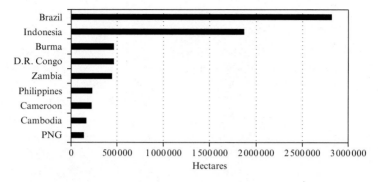

Source: FAO, *Global Forest Resources Assessment 2005* (Table 4), 196–201

Figure 5.8 Total annual deforestation in tropical countries (1990–2005)

of global *deforestation* (i.e., forestland conversion) is occurring in just three tropical hotspots: South America, Southeast Asia, and Africa. Already, a quarter of the Brazilian Amazon and half of Southeast Asia's original forests are gone. Forest degradation is also accelerating in Central and West Africa as well as within the vast northern old-growth boreal forests, particularly in Russia (which has by far the largest area of undisturbed forest among all of the 55 temperate and boreal countries).

Industrial logging operations in remote regions of the South are dislocating and impoverishing local indigenous forest-dependent communities. Companies in these places commonly acquire forestland through questionable tactics that deceive locals, leave them with few benefits, and at times even contribute to community upheaval and violence (one example, among many, is the Solomon Islands over the last two decades).[14] Destructive logging is also frequently just the beginning of a process toward complete deforestation.

From degradation to deforestation

Tropical forests are disappearing at a rate of about 12 million hectares a year. The process of deforestation generally begins after loggers damage the integrity of an old-growth forest – a common outcome in the tropics even when following "selective" or "reduced-impact" logging. A degraded forest is then more susceptible to further degradation through disturbances such as fire and infestation outbreaks. Such a forest is also easier to clear for a more commercially profitable use, such as an agricultural crop, industrial tree plantation, grazing land for livestock, or development. Most oil palm plantations in Malaysia, Indonesia, and Papua New Guinea, as well as Brazilian cattle ranches and soy plantations, occupy deforested land.

A typical cycle starts with loggers bulldozing into a pristine forest to extract high-value trees such as mahogany, lauan, or ramin. Not infrequently, as our earlier data on the illegal global wood trade shows, the timber is taken illegally. With road access now possible, and with many of the best trees now gone, others move in, commonly burning down the remaining forests.

Roads, railways, and ports provide a commodity chain portal to global markets and an incentive to harvest timber and clear forests for crops and beef. For example, the Inter-oceanic Highway in South America to traverse Peru and link with Brazil's Amazonian road network (e.g., the Trans-Amazonian highway established in the 1970s) facilitates the export of soy, beef, and other commodities via Pacific ports. Ultimately, however, they also open up previously inaccessible forest to logging, agribusiness, and mining – destroying primary intact forest and driving out indigenous communities.

Replacing these forests with industrial plantations – what some derogatorily call "green deserts" – does little to replace the environmental losses for biodiversity, air quality, water and soil integrity, or climate. Oil palm plantations support few, if any, of the species of mammals, reptiles, and birds found in primary rainforests. Timber and palm oil plantations are also encroaching on the habitat of many critically endangered species, including gorillas, elephants, orangutans, and tigers. A study of the province of Riau Indonesia by the University of Hokkaido and the WWF found that over the last 25 years, for example, Riau has lost 65 percent of its native primary forest. Of this, 25 percent was cleared for palm oil plantations, 24 percent for industrial pulpwood plantations, and 17 percent was left barren. Largely as a consequence, the study found that the Sumatran elephant population had declined by 84 percent and the Sumatran tiger population by 70 percent (over the period 1997–2007).[15]

Conclusion

Chapters 2–5 reveal a worrying overall trend for the remaining native tropical and boreal forests in the developing world. Globalizing commodity chains are shifting more and more of the forestland of the global South over to corporate interests that fuel emerging economies and feed wasteful and excessive consumption in the North. The global South is now exporting huge volumes of thousands of different low-cost wood and paper products to the North. These developing regions are also supplying increasing amounts of the world's packaging and shipping materials: from cereal and computer boxes to cargo crates and pallets. And the South is clearing more and more logged forests to establish agri-industrial plantations (especially soy and palm oil) and cattle ranches – with most of the products heading overseas to feed growing consumption in the North of meat, cheap processed food, and plant-based biofuels.

This presents a grave challenge for global sustainability:

from biodiversity to climate change. Yet, intriguingly, the rising power of big box retailers within global commodity chains is also opening up some opportunities to improve global environmental governance: the topic of our next and last chapter.

CHAPTER SIX

Governing Timber Consumption

Global commodity chains define today's political economy of timber. As this book reveals, chains linking loggers with multinational timber companies and international retailers are moving more and more of the world's timber from the tropical and boreal forests of the Third World to high-consuming states of the First World. Shiploads of consumer products are also moving from plantations and pastures established on recently deforested land of the global South to the retail shelves of the North. Economic globalization is driving – and accelerating – this process as chains lengthen and strengthen. The corporate drive into China, as every chapter in this book underscores, is particularly notable as Northern MNCs and Chinese domestic firms strive for low-cost production advantages to compete in First World markets as well as capture emerging Asian markets. One result, as we saw in the last chapter, is Russia, Southeast Asia, the South Pacific, Central Africa, and South America are increasingly bearing a disproportionate share of the social and ecological costs of consuming the world's forests.

What is being done to govern the environmental and social consequences of these corporate-led commodity chains? Answering this fully would involve analyzing a vast governance terrain. Layers and layers of rules and norms influence the particular practices of corporations within these vast chains: from state regulations to international laws to NGO campaigns to market signals. Yet imposing rules of conduct for a whole chain that criss-crosses continents is comparatively rare. States do not have jurisdictional authority to do so. NGOs do not have the resources or influence to do so. International law does not have enough power to direct the workings of the corporate world. And, until recently, corporations have not had much incentive to do so.

Over the last decade, however, the world's biggest retailers have been developing policies and programs to attempt to do precisely this – that is, monitor, change, and ultimately control the business and sustainability practices of their thousands of suppliers. Global retailers are imposing these "beyond-compliance" international requirements to move more goods faster and cheaper around the planet as well as appease NGO campaigners and reassure consumers and state regulators. Such commodity chain efforts are reinforcing growing retail authority within the corporate sphere and strengthening the capacity of retailers to not only manage risk and meet standards, but also to gain better financial and technical deals with suppliers.

The results, however, as this concluding chapter will show, are double-edged. Although these efforts are incrementally improving and raising the bar on corporate practices, at a global scale they are falling well short of tackling the forces that chapters 2–5 explained are driving forest degradation and deforestation in the global South. Moreover, these efforts are reinforcing corporate control of what is an invaluable resource for many of the world's poorest and most marginalized people – leaving it to large discount retailers to decide how, why, and where we consume timber, and further legitimizing a world of ever-increasing unsustainable consumption. Fundamentally, the attempt by big box retailers to better govern their global commodity chains raises a puzzling dilemma – can the driver of the rising discount economy really be the solution to reversing its growing negative effects?

Governing global commodity chains

Given the big box retail growth imperative to increase consumption and product sales, these companies on their own are highly unlikely to transform the current consequences of global commodity chains. To do so will certainly require changes both inside and outside the corporate realm. Internal changes include finding ways to leverage the buying and investing power of multinational companies along their global supply chains to advance sustainable production and product development and to reduce the intensity of their impacts. External ones involve harnessing market as well as state and civil society governing forces in a coordinated manner so as to bring about a broader macro-shift in how global forests and forest products are sold, processed, and valued. Critically, this shift must include regulatory enforcement that rewards responsible corporate practices while penalizing those companies that attempt to cut corners; civil society activism that advocates for greater global corporate transparency, responsibility, and accountability; and full-cost forest accounting that prices timber so as to incorporate the broader forest values of biodiversity, soil, water, and climate-regulating ecological services.

Some analysts see such "inside" and "outside" changes as mutually exclusive, labeling internal voluntary corporate social responsibility adjustments as incremental, reformist, and weak, and external regulatory changes and activist campaigns as radical and strong. We see them, in contrast, as interconnected: neither especially weak nor strong, but rather both necessary for more effective commodity chain governance. Co-regulatory governing mechanisms, involving multiple stakeholders such as governments, corporations, research institutes, and non-governmental organizations working together, tend, in our view, to make both sets of changes achievable. But these mechanisms must be well designed as co-regulation can easily fail, too, becoming in worst-case scenarios little more than a process of legitimizing corporate control, greenwash, and unsustainable timber consumption.

Keeping in mind the risks of reinforcing current patterns of unsustainable capitalism and the growing discount economy, approaches that engage and draw on both public and private resources and innovation (coordinating more traditional state-based domestic and multilateral institutional efforts with the private governance initiatives of corporations and civil society) have the potential to begin to mitigate and redistribute some of the social and environmental costs of global commodity chains – and ultimately reduce global forest loss. Co-regulatory mechanisms to govern global commodity chains are now just forming, with, as the rest of this chapter will show, big box retailers providing a potentially important contribution – breaking new ground and establishing important sustainability precedents through innovative internal supply chain greening policies and programs.

Governing from the inside

In theory, private environmental governance has the potential to go beyond the state, reaching across jurisdictions, markets, and companies with even higher standards than required by law. For sustainability, this new form of private rule-making authority is arising partly through transnational, voluntary, multistakeholder standard-setting programs. These include the United Nations Global Compact, ISO 14000, the Global Reporting Initiative, and the Forest Stewardship Council, as well as other market-based governance initiatives like the sustainable soy and palm oil roundtables. In cooperation with NGOs and mostly Northern multinational producers and manufacturers, the big box retailers are helping to develop these voluntary efforts. Private governance is also emerging through firm-level greening efforts within global commodity chains. Here, big box retailers are using their buying power to take the lead.

A directive from a single big box retailer to its thousands of suppliers can potentially have far greater reach and impact than a state regulatory effort, and much higher compliance than an international law or transnational voluntary agreement. The reason, as chapters 2–5 showed, is simple: most suppliers cannot risk a big box retailer replacing them in the commodity chain, a consequence that for smaller operators might mean immediate bankruptcy.

Indicative of the market power of these big box retailers is their growing importance as importers for Third World economies. Look at Walmart. Already, it is China's 6th largest trading partner – and purchases from here are rising as it expands its 10,000-strong network of suppliers in China and across the developing world. Its global sales – and resulting buyer power – dwarf its competitors. The 2010 Fortune 500 of America's largest corporations shows Walmart's revenues (US$408 billion) were about US$3 billion *higher* than the *combined* revenues of the following corporate titans: Costco (US$71.4 billion), Home Depot (US$66.2 billion), Target (US$65.4 billion), Lowe's (US$47.2 billion), Best Buy (US$45 billion), Amazon.com (US$24.5 billion), Staples (US$24.3 billion), Macy's (US$23.5 billion), Toys "R" Us (US$13.6 billion), Office Depot (US$12.1 billion), and Monsanto (US$11.7 billion). Walmart's buying power means it can often set terms and demand better deals from its suppliers, driving them to invest in expanded capacity and product efficiencies to meet the company's orders. A few Third World timber firms, as chapter 4 showed, are moving up the global corporate hierarchy by going global through a commodity chain: but, for many, it is leaving them caught in a cycle of increasing dependency on big box purchase orders.

> **Box 6.1   Walmart's environmental goals**
>
> 1. To be supplied 100 percent by renewable energy.
> 2. To create zero waste.
> 3. To sell products that sustain our resources and environment.
>
> *Source:* Quoted from the Walmart website, at <http://walmartstores.com/sustainability>

The greening of Walmart

Walmart is also, however, starting to use its growing influence over Third World suppliers to develop – and raise – the bar for corporate sustainability governance, not just within the retail sector but reaching into other industries, too. Walmart CEO Lee Scott's impassioned speech in October 2005, titled "Twenty-first Century Leadership," set the stage for the company's environmental efforts. Scott committed Walmart to a number of far-reaching environmental goals and targets (see box 6.1) – unprecedented commitments that one environmental writer described as "shocking the world."[1] At the time, Walmart's environmental efforts were lagging behind other global retailers, such as Home Depot and Staples. It was not so much the targets that environmentalists saw as unprecedented: rather, that these were presented as a core shift in the company's business strategy versus a peripheral add-on initiative.

Since then the company has certainly made advances. But many efforts have also stalled, faltered, and fallen well short of the CEO's ambitious vision. Much of the rollout, including initiatives to green its supply chain, has been incremental rather than momentous. For instance, Walmart's early ambition to raise Chinese production standards has stalled. The company has instead been focusing on reducing the size – and hence the freight, energy, and packaging needs – of various products sourced from China. Two examples are concentrated laundry detergent and double-roll toilet paper (reducing shipping costs and packaging).

That said, real efforts are ongoing to improve the governance of commodity chains. Walmart now has guidelines under its *Wood Furniture Supplier Preference Program* to give preference to vendors offering sustainably harvested and recycled wood products that are third-party verified. In 2008 Walmart introduced a *sustainable packaging scorecard* to encourage its suppliers to reduce the amount and environmental impacts of packaging. The following year Walmart launched a *sustainability index* to grade its suppliers and products on a range of sustainability factors (including sustainable forestry).

During this time Walmart has also supported and strengthened its supply chain greening initiatives within stakeholder partnerships. It has initiated and participates, for example, in multistakeholder "value networks," comprising government agencies, non-profits, employees, and suppliers. Also, it is taking a lead role in a global sustainability consortium – set up with academic research institutes and public- and private-sector practitioners to "build a scientific foundation to drive innovation to improve consumer product sustainability."[2] CEO Lee Scott, responding to questions at an April 2010 conference sponsored by *Fortune* magazine, acknowledged the challenges ahead, but stressed that Walmart was making progress toward his 2005 vision: "We are a huge company, and we do not always do things right. At some point, the world is going to hold corporations responsible for their environmental impact ... It's just a start, but it's a good start."[3]

Walmart has a long way to go before such efforts will start to shift the net social and ecological consequences of a growing world discount economy within which it is a major defining force. Yet these initial efforts do suggest some intriguing opportunities to draw on the market strength of multinational retail corporations to advance sustainability through global commodity chains. Specifically, the Walmart examples highlight some of the emerging supply chain business tools

Figure 6.1 Corporate supply chain greening initiatives

that leading retailers are now adopting. Many of these tools are aiming at traditional business goals, including improving efficiency, managing risk, and guarding brand reputations. At the same time these are also enhancing supply chain transparency and accountability, reducing waste and pollution, increasing availability and labeling of sustainably produced products, and decreasing raw material usage – all of which can in principle produce good outcomes for communities, consumers, and the environment. Such corporate efforts, as business author Chris Lazlo explains, create value for the company and society, with businesses ultimately, "doing well by doing good."[4]

Some of the particular initiatives, as figure 6.1 summarizes, include: supply chain tracing; forest footprint and product life-cycle assessments (including carbon accounting); supplier sustainability requirements and green procurement; eco-certification and eco-labeling; supplier audits; stakeholder partnerships; and sustainability reporting.

As we explore next, even combined, these efforts are a long way from transforming the global social and ecological consequences of retail supply chains. Still, the trend over the last decade toward more concrete corporate efforts to govern these supply chains is encouraging – and does seem to be opening up opportunities for states and civil societies to leverage these efforts to generate more significant changes.

Supply chain tracing

Being able to track products from the final customer back to their origin is a fundamental prerequisite for governing commodity chains effectively. This is not, as chapter 2 explained, a straightforward exercise for most products. This is particularly the case for composite timber products from mixed-fiber sources (for example, plywood where the high-quality outer veneer wood differs from the lower-quality core material), or for grocery products with soy and palm oil ingredients, with largely hidden connections to tropical deforestation.

The task of achieving supply chain traceability is especially challenging for the big retailers. Not only does a typical retail store sell hundreds of thousands of different products sourced from around the world, but also the store rotates products on and off its shelves as it strives to maintain product selection and offer low prices to bargain-hunting customers. Nonetheless, many are now taking up the challenge, worrying that a lack of supply chain knowledge presents a growing corporate risk: reputational, legal, and financial. In response to a protest campaign, in 2009 Walmart Brazil stopped sourcing leather and beef from Amazon farmers. In the same year, fearing reputational damage and business losses, Cadbury reversed its decision to use palm oil instead of cocoa butter in its chocolate products. And in 2010, following a highly effective and unprecedented social media campaign that used Facebook, Twitter, and YouTube to send powerful videos, images, and messages of rainforest destruction, Nestlé adopted a new policy to identify and exclude companies from its supply chain that own or

> ### Box 6.2   The Home Depot wood purchasing policy (2010)
>
> 1. The Home Depot will give preference to the purchase of wood and wood products originating from certified well-managed forests wherever feasible.
> 2. The Home Depot will eliminate the purchase of wood and wood products from endangered regions around the world.
> 3. The Home Depot will practice and promote the efficient and responsible use of wood and wood products.
> 4. The Home Depot will promote and support the development and use of alternative environmental products.
> 5. The Home Depot expects its vendors and their suppliers of wood and wood products to maintain compliance with laws and regulations pertaining to their operations and the products they manufacture.
>
> **Quoted from the Home Depot website, at <http://corporate.homedepot.com/wps/portal/Wood_Purchasing> (The Home Depot Wood Purchasing Policy has been revised several times over the past decade. Above is the policy as of October 2010.)**

manage high-risk plantations or farms linked to tropical deforestation. Retailers that lack an understanding of their wood supply chains are also now facing stiffer potential penalties for violating emerging regulatory measures to curtail the trade of illegal forest products (e.g., the Lacey Act Amendments in the United States and the EU's recently announced rules that will ban illegal timber imports).

Home Depot has been an industry leader in supply chain traceability efforts. This dates back to the 1990s campaign by environmental groups like the Rainforest Action Network to pressure the company to stop sourcing wood products from old-growth forests. The campaign was a wake-up call. Home Depot executives realized that they had little knowledge of the location, let alone the management practices, of the forests that were providing the fiber for its timber products. In response to the advocacy pressure, in 1999 Home Depot announced a commitment to avoid purchasing "endangered" wood and to source sustainably managed timber products (see box 6.2).

Home Depot then set out to map its supply chains for timber products. The company now boasts that it can trace all of its timber products: "we now know item by item – from lumber to broom handles, doors to molding and paneling to plywood – where our wood products are harvested."[5] This has enabled Home Depot to set an ambitious goal to lead its worldwide network of suppliers to understand and practice sustainable forestry. The company has also set very specific supply chain policies, including a commitment to not purchase uncertified wood products sourced from any of the 10 most vulnerable forest eco-regions identified by the WWF, and to not purchase wood products made from 40 tree species that the UN Environment Programme's World Conservation Monitoring Centre lists as potentially endangered (unless the supplier provides an export permit).[6]

Forest footprint and product life-cycle analysis

To aid in tracing and revealing how much a company's products and supply chain activities impact upon the world's forests, a number of MNCs, including several major retailers, are analyzing their "forest footprints." Some are participating, too, in an initiative launched in the United Kingdom in 2009 called the Forest Footprint Disclosure Project (or the FFD Project). A forest footprint refers here to "the total amount of deforestation caused directly or indirectly by an organization or product."[7] The FFD Project aims to increase the transparency of corporate practices within the global commodity chains of the biggest "forest risk commodities" (e.g., timber, beef, soy, palm oil, and biofuels). The financial investment community is backing the FFD Project; by 2010 there were 36 institutional investors, with assets of more than US$4 trillion, sponsoring it to help them identify which companies are managing "environmental risk" effectively. The FFD Project has two primary aims: to provide a financial incentive for companies to decrease the impact of supply chain activities; and to assist with informing consumers of the consequences for deforestation of buying different products.

The FFD Project builds on the Carbon Disclosure Project (CDP), launched 10 years earlier. The CDP, by asking companies to reveal their "carbon footprint," has created the world's biggest database of corporate climate change information. The project website explains: "2,500 organizations in some 60 countries around the world now measure and disclose their greenhouse gas emissions and climate change strategies through CDP, in order that they can set reduction targets and make performance improvements."[8]

An increasing number of companies, even in the absence of an international consensus on climate change regulation are employing carbon accounting to identify, quantify, and reduce their energy and carbon intensive processes throughout their supply chains. One example is IKEA. Introduced in their 2008 *Sustainability Report*, the company is now working to identify its carbon footprint throughout its entire value chain, "including the extraction of raw materials at source, the processing at suppliers and sub-contractors, customers transportation to IKEA stores, and customers' use of our products in their homes."[9]

Another example is Home Depot. This company is aiming to achieve continual improvements in supply chain efficiencies, including lowering greenhouse gas emissions from sourcing timber through its supply chains. As a start, Home Depot has set a target of reducing its domestic greenhouse gas emissions by 20 percent from 2008 to 2015. Walmart has also set ambitious carbon footprint targets. It is now working with the Environmental Defense Fund and others to help its suppliers to identify and quantify ways to cut greenhouse gas emissions. In February 2010 Walmart announced a plan to reduce the growth of its carbon footprint by 150 percent over the next five years – an amount equal to 20 million tonnes. Significantly, the plan is to achieve 90 percent of the company's reduction by requiring its more than 100,000 suppliers to change their practices. Tesco has also set ambitious targets to become a carbon-zero company by 2050 and to achieve a 30 percent carbon reduction in its supply chain by 2020.

More and more MNCs are also conducting broader life-cycle assessments of products and processes. Two main life-cycle assessments and global carbon accounting supply chain initiatives are the sustainability consortium (<www.sustainabilityconsortium.org>) and the World Resources Institute and the World Business Council for Sustainable Development greenhouse gas protocol reporting standards (<www.ghgprotocol.org>). This means considering the full range of impacts on the environment over the lifetime of the product at all stages, including extraction of raw materials, manufacturing, packaging, transportation, energy consumption, maintenance, and disposal.

Walmart is a leader here, creating a life-cycle analysis database of its retail stock as part of developing its sustainability index. Other retailers are actively engaging, too. Home Depot, for example, now maintains a website that provides life-cycle assessment information on the environmental impacts of its products. This is part of the company's Eco Options[SM] Program, launched in 2007 to encourage consumers to identify and purchase eco-friendly products in their stores (including sustainably produced forest and paper products).

Supplier requirements and green procurement
Similar to the Home Depot Wood Purchasing Policy, global retailers are increasingly using their purchasing power to implement green procurement policies, product specifications, and supplier sustainability requirements. Often, these go beyond legal requirements, particularly in developing

Table 6.1	Global retail company wood and paper procurement policies
Company	**Wood and paper procurement policy**
Walmart	Avoid wood from unsustainable sources, including from forests with high conservation value as well as other controversial places.
Home Depot	Give preference to wood from forests managed responsibly; eliminate wood purchases from endangered regions.
IKEA	Source all wood from verified, well-managed forests and plantations.
Lowe's	Only sell wood products that originate from well-managed, non-endangered forests.
Staples	Source paper products from "non-controversial" sources; rely on certification to ensure paper products are from well-managed forests.
Office Depot	Give preference to paper products from certified and responsibly-managed forests.

countries. Aided by sustainable procurement guides developed by groups such as the WWF, the World Resources Institute, and the World Business Council for Sustainable Development, the biggest global do-it-yourself home improvement, office supply, and hypermarket retailers all have global wood procurement policies (see table 6.1). Grocery retailers are also beginning to adopt policies to avoid the sourcing of unsustainably produced beef, soy, and palm oil products from tropical regions.

Staples, the world's largest office products company, became in 2002 the first in the paper industry to announce a comprehensive environmental paper procurement policy. "The Paper Campaign," an intensive two-year US market campaign led by ForestEthics and the Dogwood Alliance, was influential in pushing Staples to develop this policy. The policy commits Staples to only purchase "sustainable paper products" – defined as paper products "that are designed, harvested and manufactured to minimize environmental impacts across the entire life-cycle, promote responsible forest management, and protect the rights and needs of local communities."[10] Specific procurement goals include reducing demand for old-growth wood fiber, protecting forests with high conservation value, and promoting well-managed forests. The policy also had an immediate impact on competitor practices. Office Depot and Office Max, for example, followed with similar green procurement guidelines shortly after Staples announced its policy.

IKEA now has a wood procurement policy, too. Its code of conduct for its suppliers is at the forefront of such policies. Called "IWAY," it sets uniform sustainability requirements for IKEA's timber product suppliers worldwide. All of IKEA's catalogue suppliers, for example, must meet the same requirements for recycled and certified fiber. In addition, all suppliers of wood used in IKEA products must be able to document the origin of their wood (in an annual questionnaire) and, as of 2009, must perform a risk assessment of their wood supply chains. IKEA is working as well with its suppliers to encourage them to certify their operations.

Retailers are also participating in "buyer-group" initiatives to define and implement commodity chain requirements – such as green procurement policies and supplier responsibility codes – to encourage more responsible forest management and discourage or prevent the sourcing of illegal timber products. One long-standing example is the WWF-led Global Forest & Trade Network. Formed in 1991, this ties together more than 360 firms, non-governmental organizations, and communities across more than 30 countries. The goal is to expand the global market for "environmentally responsible forest products" by independently guaranteeing – and certifying – timber as legal and coming from sustainably managed forests. A second example is the European Timber Retail Coalition. Formed in 2010, this coalition of leading

European retailers – including Carrefour, Kingfisher, IKEA, and Marks & Spencer – is aiming to ensure minimum "ethical" standards for timber sold in the European Union and to reinforce the recently announced EU-wide regulations to curb illegal timber imports into the EU market.

Eco-certification and eco-labeling

Retailers are incorporating eco-certification and eco-labeled products into their commodity chain greening efforts and sales strategies as a means to define and assess supplier responsibility, ensure the sustainability of their product offerings, and capture new market demand for eco-friendly products. On the production side, eco-certification works by setting sustainability criteria that producers adopt to protect forests, water quality, wildlife habitats, local communities, and so on. On the consumption side, retailers stock their shelves with eco-certified products that are marked with an eco-label symbol to enable shoppers to pick out and purchase green products.

There are hundreds of eco-labeling programs worldwide and markets are rapidly growing. In 2009, sales of eco-products at Kingfisher alone exceeded £1 billion. Sorting through the various programs and eco-claims, however, can be a challenge for retail companies as well as the end-consumer. False green claims have become rampant. For example, a 2008–09 survey of 2,200 North American consumer products by the consulting firm EcoLogo found that more than 98 percent of the "green" products lacked proof to justify their claims. To try to stem the greenwashing tide, the US Federal Trade Commission is developing a new guide for environmental marketing claims, as well as pressing charges against some companies for deceptive labeling.

Credible eco-certification and eco-label programs are typically voluntary, involve a wide spectrum of public and private stakeholder participants in a formal decision-making body, incorporate legitimate sustainability criteria, and rely on an independent third-party accredited auditor for verification. Organizations or products found in compliance with the voluntary standard are certified – that is, the program awards it an independent "seal of approval." Organizations along the supply chain can apply for chain of custody certification. This requires them to provide a paper trail that shows how the certified product is kept separate from non-certified ones, tracked from the customer back up the supply chain to the product's point of origin. In most cases, chain of custody certification enables the product to carry the program's branded eco-label.

There are two global forest certification programs: the Forest Stewardship Council (FSC) and the Program for the Endorsement of Certification (PEFC). (The PEFC is an international umbrella program that recognizes various national-level ones.) These programs set economic, social, and environmental sustainability criteria for responsible forest management. The programs also include chain of custody certification and eco-labeling options (see box 6.3).

Over the past two decades, around 13 percent of the world's forests have been certified. To this point, however, approximately 80–90 percent has occurred in well-managed forests of Europe and North America rather than in the boreal and tropical forests of the global South. Less than 1 percent of forests in Asia and Africa, just 1.6 percent of forests in Latin America, and less than 3 percent of boreal forests in Russia are certified. The sustainable procurement policies of retailers are driving the growth in certified product demand; yet, so far, eco-certified forest products have had limited impact on mitigating tropical or boreal deforestation in the developing world.

Many factors explain the lack of forest certification in the global South. These include high financial costs, limited

> **Box 6.3   Forest certification**
>
> Forest certification emerged in the early 1990s out of growing public awareness and concerns about deforestation, and out of frustration with the stalled international attempts to establish a global forest convention to protect the world's remaining old-growth forests. Advocates saw certification as a way to fill a gap in global forest governance, create a market incentive for continuous improvements in forest management, and provide a means for consumers to identify and shift their purchases toward legal and sustainably managed forest products.
>
> Retailers such as Home Depot and B&Q were key players in the establishment of the Forest Stewardship Council in 1993, alongside environmental groups such as Friends of the Earth and the WWF. The large Northern integrated forest companies led the development of the biggest national programs: the Sustainable Forestry Initiative in the United States (1995) and the CSA Z809 standard in Canada (1996). A few years later (1999) small family forest owners in the Nordic countries established the Program for the Endorsement of Certification (before it was turned into an international umbrella program in 2003).
>
> While various groups continue to this day to debate the strengths and weaknesses of the two global programs, retailers generally recognize both standards as legitimate and credible. Some retailers state a preference for the FSC as it has the greatest acceptance among environmental advocacy organizations. For example, in 2009, the UK do-it-yourself retailers Focus and Homebase purchased around two thirds of their wood from FSC-certified sources.

governance capacity, poorly defined property rights, lack of coordination among many small producers, and weak market signals. Recently, however, forest certification adoption is slowly growing in developing countries as certification awareness and market demand increases, as landowners and forest producers organize better, and as governments begin to subsidize certification. Yet problems abound. Both APRIL and Sinar Mas have recently had their FSC forest certifications revoked for violating requirements of the standard, such as the prohibitions against converting tropical rainforests into timber plantations.

Despite the challenges, forest certification continues to gain a foothold in the global timber markets. Similar to forest certification, eco-certification programs are also under development to identify responsibly produced soy, palm oil, beef, and biofuels. Retailers, as well as the large multinational forestry and agro-food corporations, are all now participating on the decision-making bodies of these various multistakeholder eco-certification initiatives, as well as working to eco-certify their own operations.

Supplier audits

Many retailers are also starting to audit suppliers to verify that they are meeting certification standards, corporate codes of conduct, and commodity chain sustainability requirements. IKEA now has a Compliance & Monitoring Group, for example, that follows up on its IWAY code with regular audits (in addition to its third-party audits conducted by the Rainforest Alliance SmartWood Program). In China, in particular, IKEA audits every supplier at least once a year – with most audit visits now *unannounced*. Suppliers found in non-compliance are given a chance to take corrective actions, with IKEA introducing *IWAY developers* to provide training and advice to help improve conditions in Chinese factories. If further instances of non-compliance occur, however, IKEA then terminates the business relationship. IKEA is not yet auditing all of its wood supply chain; but of the portion audited, 92 percent was already meeting the company's forestry requirements by 2009.[11]

Recognizing the challenge of auditing thousands of suppliers within vast and complex global commodity chain pathways, many retailers are now working with certification organizations – as well as partnering with NGOs – to widen the coverage of their supplier audit program, particularly within sensitive or controversial forest regions. Office Depot, for example, has partnered with Conservation International and the Nature Conservancy to audit supplier forest sustainability practices in the Atlantic forest region of the southern Brazilian state of São Paulo – where large areas of native forest have been converted into plantations.

Stakeholder partnerships

Retailers are increasingly participating in strategic partnerships among multiple stakeholders to design, implement, and legitimize policies and programs for governing the sustainability of global commodity chains. These partnerships involve conservation groups (e.g., WWF, Conservation International, and the Nature Conservancy), buyer groups (e.g., the WWF Global Forest & Trade Network), green building associations (e.g., US Green Building Council), and business councils (e.g., World Business Council for Sustainable Development).

IKEA's Wood Procurement Group, for example, has not only established long-term agreements with governments, forest owners, and forestry-related associations, but also formed a forest cooperation partnership with the WWF to encourage responsible forestry, discourage illegal logging, and promote certification. As well, since 2005, IKEA has been working in partnership with the Rainforest Alliance to improve forest practices and increase forest certification awareness in China. The Office Depot's 2008 *Corporate Citizen Report*, meanwhile, explains how the company has provided US$2.2 million of financial support for sustainable forestry as part of an alliance with Conservation International, NatureServe, and the Nature Conservancy. Home Depot has also given more than US$1.5 million to environmental non-profit groups and partnered with the Nature Conservancy to promote sustainable timber harvesting in Indonesia. The company has partnered as well with environmental groups, governments, and industry in the Brazilian Amazon to educate the local communities about sustainable timber harvesting.

Environmental groups have also proactively supported

and rewarded positive corporate sustainability initiatives. The Natural Resources Defense Council (NRDC), for example, strongly backed the decision by Staples to source paper certified by the Forest Stewardship Council as containing post-consumer recycled content. "Staples' leadership," explains Debbie Hammel, Senior Resource Specialist with the NRDC, "sends a loud signal to other corporate consumers and paper producers that the market demands strong environmental performance standards."[12]

Sustainability reporting
Finally, recognizing the increasing demand from governments and consumers for greater multinational transparency and accountability, the global retailers are communicating their sustainability commitments and global commodity chain greening programs through sustainability reports that document their progress toward sustainability goals. Some are using the Global Reporting Initiative (GRI) guidelines to facilitate benchmarking of their performance against the progress of other companies. Some are also hiring professional auditing firms like PricewaterhouseCoopers to conduct independent assurance assessments of the sustainability reports, so as to verify the legitimacy of the reported information and avoid accusations of corporate greenwash. Others are participating in retail industry voluntary codes of conduct, such as Europe's *Retail Environmental Sustainability Code*, in which signatory companies agree to adopt better environmental management practices and to track and report regularly on their sustainability progress.

The prospects and limits of retail power

There is, then, a wide range and growing number of private corporate initiatives to disclose, mitigate, and manage the social and environmental impacts of global commodity chains. Measurable changes are occurring. The amount of packaging for some products is going down. Recycling is rising. Procurement of timber products is becoming more responsible among firms with enforced policies to shift wood purchases toward legal and sustainably produced products. More than 80 percent of B&Q's timber supplies, for example, now come from certified legal and well-managed sources, while all hardwood garden furniture sold at Asda is FSC-certified. And, lastly, certified forest product markets are gradually expanding, particularly for eco-labeled tissue and office products with post-consumer recycled content. For example, Tesco's privately branded toilet tissue, kitchen towel, and face tissues are all FSC-certified.

Nonetheless, the overall effectiveness, especially for the sustainability of tropical forests of the global South, is still falling well short as the consequences of rising discount merchandizing by big box retailers overwhelm the incremental advances of the new policies and programs. As highlighted earlier, voluntary global programs such as forest certification are failing to achieve significant results in the tropics. Less than 1 percent of tropical forests have been certified; and certified forest products account for less than 10 percent of the global timber market. Moreover, consumers (particularly those in North America) remain largely unaware of certification eco-labels or the impact of their timber and grocery product purchases on deforestation.[13]

Companies, too, are not reaching some of their targets. IKEA is working with partners in high-risk supply chain areas such as Russia and China to increase the growth of third-party certified fiber and implement chain of custody certificates among IKEA suppliers. But progress has been slow. The company is not yet close to its target of a 30 percent share of the wood used in IKEA products coming from forests certified as responsibly managed – as of mid-2010 it was at just 7 percent.

Walmart is also falling well short of its ambitious sustainability targets. Its goals of running on 100 percent renewable energy, producing zero waste, and selling products that sustain the global environment raise many skeptical eyebrows. In response, the company describes these as "aspirational" – but that it is stepping steadily toward them by, for example, designing more energy-efficient stores, increasing the efficiency of its truck fleet, and reducing packaging waste.

Critics nonetheless see Walmart as essentially ignoring its escalating impact from a business growth strategy that relies on increasing global consumption. Michael Brune, the executive director of the Sierra Club, explains: "Wal-Mart right now is in a spot where there is legitimately a lot of praise, but there is a lot to legitimately criticize. The scale of their effort is grossly surpassed by the pace of the company's growth."[14] Tellingly, the company's greenhouse gas commitment is to cut the *intensity* of its total carbon footprint – not its *total* emissions.

Together, then, the sustainability policies of big retailers are incrementally – yet steadily – reducing the per unit intensity of the activities of suppliers within global commodity chains. This is a notable accomplishment. And the growing power of big box retailers within these chains suggests that the capacity of such policies to influence corporate practices *globally* will have a multiplier effect – expanding and continuing to increase in the future as they raise benchmarks of responsible business practices across industry sectors and countries. For states, international organizations, and civil society groups this presents a significant opportunity to leverage this growing power and the emerging sustainability policies to advance global environmental governance through co-regulatory and market-led mechanisms. Still, this alone will never suffice to overcome the profit-logic of global commodity chains. Broader, systemic changes are also vital.

Going beyond corporate social responsibility

Globally, it is critical to address the inequality of progress toward more sustainable practices – a process that is deforesting the global South to produce cheap timber, food, animal feed, and biofuel for export to the North. This will require far better co-regulatory coordination among international state-led mechanisms, civil society advocacy, voluntary standard setting, and global commodity chain market strategies. On the production side, beyond better regulatory enforcement, strong advocacy pressure and greater certification subsidies in developing countries are necessary. Governments and industry will also need to implement full-cost accounting with the aim of internalizing the social and environmental value of forests into timber prices and land-use decisions. On the consumption side, to help stem the tide of wasteful, discount-driven consumerism, credible eco-labeling programs and green product market demand will need to expand. As well, the eco-consumer discourse will need to become far stronger to counter the message of the big box retailers to buy more bargains. "What's good for Walmart is *not always* good for the planet."

The 20 years of failed efforts to negotiate an international agreement for global forest governance shows the difficulty of finding solutions to deforestation and the ecological and social costs of timber consumption. Transforming the social and ecological consequences of globalizing timber chains will require far-reaching change. The prices of goods traded within these chains – as our analysis in chapters 2–5 demonstrated – largely ignore the environmental and social costs, from the loss of biodiversity to the emission of carbon dioxide, to the erosion of community prosperity and well-being. Powerful forces support this. Multinational investors and international development agencies provide essential financing. And governments provide attractive subsidies and incentives, such as low fees and tax breaks.

Third World forest and paper companies within these global commodity chains are under intense pressure to maintain low production costs to sustain access to Northern markets. This is particularly true for those firms supplying market channels that are narrowing to retail buyers like Walmart or Home Depot or Lowe's. Third World suppliers cannot afford, as we saw in chapter 4, to lose their position within these chains. Good quality and consistency are crucial: but it is even more important to keep costs as low as possible to avoid the big buyers switching to a supplier providing the product for less. Suppliers, especially ones operating in states with weak regulatory capacity, end up competing hard to keep prices as low as possible: doing everything from logging illegally to bribing officials to disregarding environmental safeguards.

Corporations end up mining forests for timber – and then clearing them for industrial plantations and ranches – to line row after row of ever more powerful big box retailers with bargain-priced goods. The list of such products could go on for many pages: easy-to-assemble furniture, wooden toys, flooring, picture frames, tissue, fine paper, disposable diapers, leather purses, cosmetics, chocolate bars, steaks, and so on.

Shipping and trucking all of this – and much more – to Northern consumers relies as well on the forests of the global South for the packaging. About one quarter of the world's timber now goes into items such as pallets, boxes, and wrapping for everything from food to sofas to computers. All of this is allowing big retailers like Europe's Carrefour and America's Walmart to prosper: but it is also stimulating wasteful and excessive consumption as buyers do not pay the full costs for extracting, processing, and transporting these bargain-priced products.

The growing power of Northern corporations – and to a lesser extent Southern firms – over the world's forests is driving and reinforcing these trends: not necessarily by fencing out locals or by coercing state leaders, but by defining and controlling the value of forests as commercial timber, where most of the financial profits go into the coffers of big business and most of the ecological and social costs go into poor communities and fragile environments. This process is not set in stone, however. Corporate power within commodity chains, as we revealed in chapters 2–4, is highly dynamic, with firms rising and falling as markets and politics shift. Some Third World timber firms, as we documented in chapter 4, are growing rapidly after going global and linking into a global commodity chain – often at a much faster rate than the trajectories of the older Northern multinational powerhouses.

Yet the top Northern MNCs still retain far more power, with no Third World firm yet in the global top 10. The Northern MNCs continue to invest in large timber operations and manufacturing facilities around the world. They continue to supply the large publishing, shipping, and construction sectors. And, increasingly, they are branding forest and paper products for direct retail sale. Nevertheless, as our analysis shows, the traditional power of Northern integrated timber companies within the corporate sphere – from legal contracts to policy discourse – is diffusing: not so much to Third World firms that generally depend on Northern financing, markets, and contacts, but to big box retailers that are controlling the flow of more and more of the world's traded goods.

Such trends, as this book shows time and again, are contributing to some highly problematic consequences, including destructive logging and deforestation in the global South, with relatively little income or long-term benefits for local peoples. At the same time, however, as we have seen in this final chapter, shifts in power within the corporate sphere may also be opening up some opportunities to improve global forest governance. Most significantly, the capacity of big retail stores to impose sustainability policies on a *global* scale is rising as more Third World timber firms go global and as suppliers and investors increasingly depend on these retailers for profits and financial stability. A store like Walmart can now arguably wield more power than a state over a logging or timber processing company – especially within the weaker states of the global South. Big box retailers recognize this. And, as this chapter documented, all of them have been developing new green supply chain policies and codes of conduct, motivated in part to reassure consumers and retain markets and in part to increase their authority and market control within global commodity chains.

Are these voluntary greening policies and programs improving corporate practices and global forest governance? The results so far, as our conclusion shows, are measurable – but incremental – and are still more about changing the discourse rather than the actions of firms. They are not nearly enough to offset the broader effects of big box stores as engines of consumption driving a growing world discount economy. Still, our analysis does reveal some intriguing co-regulatory prospects to promote global sustainability and more responsible business practices – most notably if states and civil societies can tap into, and then leverage, the growing power of the retailers to act as responsible sustainability leaders within globalizing commodity chains. But supporting private environmental governance and pursuing co-regulation also involves a real danger. It can just as easily end up legitimizing the authority of large multinationals in ways that accelerate the already fast-growing world discount economy, aggravate existing South–North inequities, and further increase rising rates of unsustainable consumption: with an outcome of worse – not better – global environmental governance.

Notes and References

CHAPTER I: THE GLOBAL POLITICAL ECONOMY OF TIMBER

1 Internationally, analysts and policymakers generally classify forests along four lines that distinguish the type of forest by the management conditions (primary, secondary, plantation); geographic location (tropical, temperate, boreal); species (deciduous, coniferous); and type of fiber (hardwood, softwood). Although not fully accurate scientifically, various categories are commonly combined into two broad groupings:

Tropical	Boreal
Deciduous	Coniferous
Hardwood	Softwood

Tropical (and sub-tropical) forests comprise more than half of the world's forests. Most are still primary (old-growth) forests, meaning that over the last few hundred years these have not been logged or naturally disturbed in a major way (such as by an insect infestation, windstorm, or fire). Located close to the equator in warm humid climates, these mostly comprise hardwood deciduous species. Those in the Brazilian Amazon, African Congo, and Southeast Asia are the largest, with as many as 1,000 different tree species, including commercially prized teak and mahogany. Tropical forests are unique, not only because of their rich biodiversity (containing something like 70 percent of the world's plants and animals), but also because the tree biomass (rather than the soil) holds most of the forest nutrients. Thus, unlike temperate or boreal forests, tropical ones lose soil fertility very quickly after harvesting, making regeneration a greater challenge.

Boreal forests, which stretch in a belt across northern Canada and Eurasia (Scotland, Norway, Sweden, Finland, and Russia), account for one third of global forest area. Sometimes called "taiga," boreal forests thrive in cold climates, with fewer species and lower productivity than tropical forests. Softwood conifer species such as pine, spruce, and fir dominate the boreal forests. Boreal forests in Western Europe and Russia are mostly secondary forests, while the majority in Canada are primary ones.

Temperate forests cover a comparatively small area, accounting for just over one tenth of the total. Found mainly in East Asia (China), Europe, eastern Canada, and the US, temperate forests include a mix of deciduous (e.g., oak, maple, beech, cherry) and conifer trees, and thus produce both hardwood and softwood fiber. Most are secondary forests: logged and reforested largely with faster-growing conifers. One significant exception is the oceanic temperate rainforests of North America's Pacific Northwest. These have the largest, tallest, and oldest trees in the world, including: "General Sherman," a Sequoia tree in California with a 31 meter trunk, a height of 84 meters, and an age of 2,300 to 2,700 years; and the "Mendocino Tree," a 1,000-year-old Coastal Redwood reaching over 110 meters (and still growing).

2   We should stress that achieving accurate assessments of deforestation and timber flows is an ongoing challenge. Illegal logging and timber smuggling are common. So is misreporting of volumes and species to lower taxes, fees, and tariffs. Governments define and measure forests differently, too. Some forest surveys, for example, include industrial timber plantations but exclude small woodland patches or areas without commercial tree species. The United Nation's Food and Agriculture Organization (FAO) conducts the global forest inventory and defines a forest as including, "a minimum threshold for the height of trees (5 m), at least 10% crown cover and a minimum forest area of 0.5 hectares." However, these definitional thresholds for forest height, crown, and area differ among countries. Combining remote sensing satellite imagery with country reports nonetheless does reveal some reasonably stable estimates and trends.

Predictably, then, estimates of the contribution of deforestation to global carbon dioxide emissions vary somewhat, with most analysts settling on about one fifth of the total (including the

Intergovernmental Panel on Climate Change and the 2006 *Stern Review of the Economics of Climate Change*). One recent study, however, put the figure at just 12 percent (with a further 3 percent from tropical peatlands). See Guido R. van der Werf et al., "CO_2 Emissions from Forest Loss," *Nature Geoscience* 2 (11) (2009): 737–8.

3 For an elaboration of the "shadows of consumption," see Peter Dauvergne, *The Shadows of Consumption: Consequences for the Global Environment* (Cambridge, MA: MIT Press, 2008).

4 FAO, *Global Forest Resources Assessment 2010: Key Findings* (Rome: FAO, 2010), 19; Gerald Urquhart et al., "Tropical Deforestation," 2001, Reference for Earth Observatory NASA website at: <http://earthobservatory.nasa.gov>.

5 FAO, *State of the World's Forests* (Rome: FAO, 2009), 82.

6 A significant contribution to understanding buyer- versus seller-driven commodity chains was Gary Gereffi's chapter, "The Organization of Buyer-driven Global Commodity Chains: How US Retailers Shape Overseas Production Networks," in a book he edited with Miguel Korzeniewicz, *Commodity Chains and Global Capitalism* (Westport, CT: Praeger, 1994). Other examples from this literature are Gary Gereffi, John Humphrey, and Timothy Sturgeon, "The Governance of Global Value Chains," *Review of International Political Economy* 12 (1) (2005): 78–104; Jennifer Bair, "Global Capitalism and Commodity Chains: Looking Back, Going Forward," *Competition & Change* 9 (2) (2005): 153–80; and Jennifer Bair, ed., *Frontiers of Commodity Chain Research* (San Francisco, CA: Stanford University Press, 2009).

7 This understanding of power draws on Doris Fuchs's typology of instrumental, structural, and discursive corporate power (which we use explicitly in Peter Dauvergne and Jane Lister, "The Power of Big Box Retail in Global Environmental Governance: Bringing Commodity Chains Back into IR," *Millennium* 39 (1) (2010): 145–60. For a full explanation of the typology, see Doris Fuchs, "Commanding Heights? The Strength and Fragility of Business Power in Global Politics," *Millennium: Journal of International Studies* 33 (3) (2005): 771–801. Also, see her book, Doris Fuchs, *Business Power in Global Governance* (Boulder, CO: Lynne Rienner, 2007). Her framework drew on David Levy and Daniel Egan, "Corporate Political Action in the Global Polity: National and Transnational Strategies in the Climate Change

Negotiations," 138–53, in Richard Higgott, Geoffrey D. Underhill, and Andreas Bieler, eds., *Non-State Actors and Authority in the Global System* (London: Routledge, 2000). Fuchs later joined with Jennifer Clapp in a pathbreaking edited book using this typology: Jennifer Clapp and Doris Fuchs, eds., *Corporate Power in Global Agrifood Governance* (Cambridge, MA: MIT Press, 2009). Within this book, for an analysis of the growing concentration of retail power – especially structural and discursive – see chapter 2, Doris Fuchs, Agni Kalfagianni, and Maarten Arentsen, "Retail Power, Private Standards, and Sustainability in the Global Food System," 30–59.

Also, see Peter Newell, "The Marketization of Global Environmental Governance: Manifestations and Implications," 77–94, in Jacob Park, Ken Conca, and Matthias Finger, eds., *The Crisis of Global Environmental Governance: Towards a New Political Economy of Sustainability* (London: Routledge, 2008); Matthew Paterson, *Automobile Politics: Ecology and Cultural Political Economy* (Cambridge, UK: Cambridge University Press, 2007); and Robert Falkner, *Business Power and Conflict in International Environmental Politics* (London: Palgrave Macmillan, 2007).

8 Quoted in Steven Mufson, "In China, Wal-Mart Presses Suppliers on Labor, Environmental Standards," *Washington Post*, 28 February 2010, G01. Also, for a discussion of the importance of Walmart for supplier markets, see Kris Hudson, "Wal-Mart Ripple Effect Strikes Again: Cutbacks Weigh on Supplier Earnings," *The Wall Street Journal Online*, April 27, 2006, at: <www.reveredata.com/about/news/press/20060427.html>.

9 The characteristics of particular commodity chains vary by product, region, and corporate arrangements. Significant differences exist even within seemingly similar industry segments: for example, between kitchen cabinets, wooden desks, and wooden chairs. Variations can arise across short timeframes as corporations respond to things like market shocks, political turmoil, or mergers and acquisitions. Our book attempts to uncover general patterns across and within timber supply chains, but still recognize the great diversity of particular chains. Chapter 2 explores this further.

10 The forest industry refers to trees that have been cut down and removed from a forest as *roundwood*. This includes both woodfuel and industrial roundwood. Each year about 3.6 billion m3 of

roundwood is produced, which is equal to about 1.5 percent of the global timber reserve. Around half of all roundwood is used for industrial purposes, with sawlogs and pulplogs becoming panels, lumber, paper, and paperboard – or, for simplicity, what we call "timber." The other half is used as fuel largely for heating and cooking. The percentage of wood used for fuel is much higher in tropical countries (over 80 percent on average) than in temperate ones (about 25 percent on average). Actual woodfuel removals are hard to estimate, however, because most of the collection is informal – armloads of branches and logs gathered by local people.

11 Indonesia, for example, restricted raw log exports in the 1980s to stimulate domestic processing, and quickly became the world's leading tropical plywood exporter. More recently, in February 2007, Russia announced an 80 percent raw log export tariff in order to encourage domestic processing capacity (although implementation has been subsequently delayed). Gabon also recently banned log exports, effective January 2010.

12 For the estimate on the amount of the world's traded timber going through China, see William F. Laurance et al., "The Need to Cut China's Illegal Timber Imports," *Science Magazine* 29 (February) (2008): 1184–5. For a database of reports and statistics, see the Chatham House illegal logging website at <www.illegal-logging.info>.

13 Over the last decade or so, scholars of global environmental politics have been uncovering a pattern of growing corporate power to set agendas and influence norms, regimes, and institutions. A few examples are Graeme Auld, Steven Bernstein, and Benjamin Cashore, "The New Corporate Social Responsibility," *Annual Review of Environment and Resources* 33 (2008): 413–35; Peter Newell, "The Political Economy of Global Environmental Governance," *Review of International Studies* 34 (3) (2008): 507–29; Matthew Paterson, "Global Governance for Sustainable Capitalism? The Political Economy of Global Environmental Governance," 99–122, in Neil Adger and Andrew Jordan, eds., *Governing Sustainability* (Cambridge, UK: Cambridge University Press, 2009); Doris Fuchs, *Business Power in Global Governance* (Boulder, CO: Lynne Rienner, 2007); Robert Falkner, *Business Power and Conflict in International Environmental Politics* (London: Palgrave Macmillan, 2007); Jennifer Clapp, "Global Environmental Governance for Corporate

Responsibility and Accountability," *Global Environmental Politics* 5 (3) (2005): 23–34; David Levy and Peter Newell, eds., *The Business of Global Environmental Governance* (Cambridge, MA: MIT Press, 2005); Benjamin Cashore, "Legitimacy and the Privatization of Environmental Governance: How Non-State Market-Driven (NSMD) Governance Systems Gain Rule-Making Authority," *Governance* 15 (4) (2002): 503–29; Leslie Sklair, "The Transnational Capitalist Class and Global Politics: Deconstructing the Corporate-State Connection," *International Political Science Review* 23 (2) (2002): 159–74; Rodney Bruce Hall and Thomas J. Biersteker, *The Emergence of Private Authority in Global Governance* (Cambridge, UK: Cambridge University Press, 2002).

14 The estimate of Indonesia's greenhouse gas emissions was made by PEACE (Pelangi Energi Abadi Citra Enviro), *Indonesia and Climate Change: Current Status and Policies* (Jakarta: PEACE, 2007).

CHAPTER 2: THE POWER OF BIG RETAIL

1 Michael Porter, *Competitive Strategy: Techniques for Analyzing Industries and Companies* (New York: Free Press, 1989); Michael Porter, *Competitive Advantage: Creating and Sustaining Superior Performance* (New York: Free Press, 1985).

2 See Gerard Cliquet, "Large Format Retailers: A French Tradition Despite Reactions," *Journal of Retailing and Consumer Services* 7 (4) (2000): 183–95; Gordon Laird, *The Price of a Bargain* (Toronto: McClelland & Stewart, 2009), 22; Stephen Arnold and Monika N. Luthra, "Market Entry Effects of Large Format Retailers: A Stakeholder Analysis," *International Journal of Retail & Distribution Management* 28 (4/5) (2000): 139–54.

3 The statistics in this paragraph are from: Emily Matthews, Richard Payne, Mark Rohweder, and Siobhan Murray, *Pilot Analysis of Global Ecosystems* (Washington, DC: World Resources Institute, 2000); Resource Conservation Alliance, *Wood Consumption*, <www.woodconsumption.org>; Forrester Research Inc. US *Online Retail Forecast 2009–2014*, March 8, 2010; The United Nations System-wide Earthwatch at <http://earthwatch.unep.net/emergingissues/consumption/ reducconsump.php>;

Deloitte, *Global Powers of Retailing 2010* (London: Deloitte Touche Tohmatsu).

4 The statistics in this paragraph are from: Walmart corporate website; Pew Research Center, "Wal-Mart – A Good Place to Shop But Some Critics Too," *Survey Report,* Pew Research Center for the People & the Press, Washington, DC, December 15, 2005; Suzanne Kapner, "Wal-Mart Celebrates its Growing Market Share," *CNNMoney.com*, June 5, 2009; Andrea Felsted, "Market Share Slip Forces Asda to Fight Back," *FT.com (Financial Times* online), April 28, 2010.

5 Eswar Prasad, ed., *China's Growth and Integration into the World Economy: Prospects and Challenges,* International Monetary Fund, Occasional Paper 232, Washington, DC, 2004, p. 51. As well, for a more recent analysis of the growing role of emerging economies like China and India in driving economic growth and innovation, see "A Special Report on Innovation in Emerging Markets," *Economist*, 15 April 2010.

6 Quoted in Daniel Zoll, "Opposition Growing to Home Depot," *Albion Monitor,* March 30, 1997, at: <www.albionmonitor.com/9703b/homedepot.html>.

7 Constanza Bianchi, "Home Depot in Chile: Case Study," *Journal of Business Research* 59 (3) (2006): 391–3.

8 Other terms for commodity chain include "supply chain" and Michael Porter's "value chain," which refers to the gains in value as the product passes through each stage in the production chain – with the overall value greater than the sum of the parts. In this book we use the overlapping business terms "supply chain" and "value chain" interchangeably with the term "commodity chain." For an overview of the literature, see Jennifer Bair, ed., *Frontiers of Commodity Chain Research* (San Francisco, CA: Stanford University Press, 2009); also see the Global Value Chain initiative at Duke University at <www.globalvaluechains.org>.

9 Raphael Kaplinsky, Olga Memedovic, Mike Morris, and Jeff Readman, *The Global Wood Furniture Value Chain: What Prospects for Upgrading by Developing Countries* (Vienna: United Nations Industrial Development Organization, 2003); Herry Purnomo, Philippe Guizol, and Dwi R. Muhtaman, *Governing Teak Furniture Business: A Global Value-Chain System Dynamic Modeling Approach* (Bogor, Indonesia: Center for International Forestry Research, 2007).

10 Andy White et al., *China and the Global Market for Forest Products: Transforming Trade to Benefit Forests and Livelihoods* (Washington, DC: Forest Trends, March 2006), 16.

11 Buyer power refers to "the ability of one or more buyers, based on their economic importance on the market in question, to obtain favourable purchasing terms from their suppliers." See European Union DG Competition Glossary of Terms, 2003, at <www.concurrences.com/rubrique.php3?id_rubrique=161>. As well, see Gary Gereffi, "The Organization of Buyer-Driven Global Commodity Chains: How US Retailers Shape Overseas Production Networks," in Gary Gereffi and Miguel Korzeniewicz eds., *Commodity Chains and Global Capitalism* (London: Praeger, 1994), 95–122.

12 "Upgrading" refers to the process of adding value to products by improving their quality, making them more efficiently, or moving into more skilled activities and sophisticated product lines. For foundational literature on commodity chain upgrading, see Gary Gereffi, "International Trade and Industrial Upgrading in the Apparel Commodity Chain," *Journal of International Economics* 48 (1) (1999): 37–70; John Humphrey and Hubert Schmitz, *Developing Country Firms in the World Economy: Governance and Upgrading in Global Value Chains*, INEF Report, No. 61 (Duisburg: University of Duisburg, 2002); Michael Porter, *The Competitive Advantage of Nations* (London: Macmillan, 1990). For a more recent study on upgrading, see Carlo Pietrobelli and Roberta Rabellotti, *Upgrading to Compete: Global Value Chains, Clusters, and SMEs in Latin America* (Boston, MA: Harvard University Press, 2007).

13 See Charles Fishman, *The Wal-Mart Effect: How the World's Most Powerful Company Really Works – and How It's Transforming the American Economy* (New York: Penguin, 2006); Charles Fishman, "The Wal-Mart You Don't Know," on the website Fast Company (posted December 1, 2003), at <www.fastcompany. com/magazine/77/walmart.html>. For more background on Walmart, see Ellen Ruppel Shell, *Cheap: The High Cost of Discount Culture* (New York: Penguin, 2009); Gordon Faircloud, "Wal-Mart Sneezes, China Catches Cold," *Wall Street Journal*, May 29, 2007, B1–2; Stacy Mitchell, *Big-Box Swindle* (Boston, MA: Beacon Press, 2006); Stanley D. Brunn, ed., *Wal-Mart World: The World's Biggest Corporation in the Global Economy* (London: Routledge, 2006).

14 In particular, the Environmental Protection Agency traced the origin of the 200,000 poplar and birch wooden baby cribs provided to Walmart each year by Dalian Huafeng Furniture Co., one of China's largest manufacturers of secondary wood products. They found that the main wood supplier to the Huafeng mill was the Russian company, Chuguevskaya LPK. This company was logging as well as buying its wood (from companies such as Araliya Ltd and Svetlyanka Ltd) in areas with some of the highest illegal logging rates in Russia as well as in the middle of an endangered tiger breeding habit. Further, the EIA investigators observed how the wood for the cribs was being sourced through Suifenhe city, a major Chinese gateway on the Russian border where protection payments to Russian mafia were common. See Environmental Investigation Agency, *Attention Wal-Mart Shoppers: How Wal-Mart's Sourcing Practices Encourage Illegal Logging and Threaten Endangered Species* (Washington, DC: Environmental Investigation Agency, 2007). The Environmental Investigation Agency's 2006 study is EIA/Telapak, *Behind the Veneer: How Indonesia's Last Rainforests are being Felled for Flooring* (Washington, DC: Environmental Investigation Agency, 2006).

As one indicator of China's timber sources, in 2005 the country imported approximately 70 percent of its timber products from Russia (49 percent), Indonesia (5.7 percent), Malaysia (8.3 percent), Thailand (4.6 percent), and Papua New Guinea (4.2 percent). See White et al., *China and the Global Market*, 12. As of 2005, the percentage of total timber production in these countries from illegal sources was estimated at: Russia (10–50 percent); Indonesia (70–80 percent); Malaysia (35 percent); Thailand (40 percent); and Papua New Guinea (70 percent). For a summary of estimates, see: The World Bank, *Strengthening Forest Law Enforcement and Governance* (Washington, DC: World Bank, August 2006), 9.

CHAPTER 3: THE NORTHERN FOREST AND PAPER MULTINATIONALS

1 Interview with Annie Zhu, *Pulp & Paper International (PPI) Magazine*, July 30, 2008, on the Resource Information Systems

Inc. (RISI) website, at <www.risiinfo.com/technologyarchives/
PPIMagJuly-Biggest-achievement-people-development.html>.

2 PricewaterhouseCoopers (PwC), *Global Forest, Paper & Packaging
Industry Survey: 2009 Edition* (Vancouver, BC: PwC LLP, 2009), 9.
The PricewaterhouseCoopers global ranking relies on compiled
company data. The analysis excludes several of the top 100 private
forest companies because of the lack of publicly available financial
information. These companies include: Georgia-Pacific, Kruger,
Tolko, and Menasha in North America; Tetrapak, SAICA, Fritz
Egger, Elopak, Papierfabrik Palm, Fedrigoni Group, and Ilim
Pulp in Europe; and Asia Pulp and Paper, APRIL, Rank Group,
and Visy Industries in the Asia Pacific.

3 J. A. Lamberg, J. Näsi, J. Ojala, and P. Sajasalo, eds., *The Evolution
of Competitive Strategies in Global Forestry Industries: Comparative
Perspectives* (Dordrecht, The Netherlands: Springer, 2006).

4 For example, over the last decade and a half, many of the large
American integrated timber companies such as International
Paper have sold off a large share of their private timberland
holding. As a consequence, a new form of corporate timber
owner has been emerging in the United States. Institutional
timber investment organizations (TIMOs) and real estate
investment trusts (REITs) have taken over more than half of the
private industrial forestland. For example, see: The Campbell
Group at <www.campbellgroup.com/about/index.aspx>.

5 PricewaterhouseCoopers, *Branching Out: Global Deal Activity in
the Forest, Paper & Packaging Industry, 2008 Edition* (Vancouver,
BC: PwC LLP), 2–3.

6 Jason Bush, "Now It's Really International Paper," *Business Week*,
December 17 (2007), 63; Chris Lang, *Banks, Pulp and People: A
Primer on Upcoming International Pulp Projects*. A report prepared
for Urgewald (Sassenberg, Germany: Urgewald, 2007), 13. On
fiber costs, see PricewaterhouseCoopers, *Global Forest*; and for
global wood price data, see *Wood Resource Quarterly* at <www.
woodprices.com>.

7 An exception is India, where the government has not provided
privileged land access to the pulp and paper companies. Industry
analysts argue that this has contributed to India lagging behind
China, Indonesia, and Brazil in the global forest and paper sector.
None of the major global forest companies, for example, had set
up paper-making capacity in India as of early 2010.

8 The case of the company Veracel is from Lang, *Banks, Pulp and People*, 6–7; and the Veracel company website at <www.veracel. com.br/en>. For an analysis of the role of debt and structural adjustment on deforestation, see John M. Shandra, Eran Shor, Gary Maynard, and Bruce London, "Debt, Structural Adjustment, and Deforestation: A Cross-National Study," *Journal of World-Systems Research* XIV (1) (2008): 1–21.

9 Bush, "Now It's Really International Paper," 63.

10 Jan Johansson, SCA President and CEO, SCA Company presentation at Handelsbanken Conference, September 21, 2009, 20–1, at <www.sca.com/Documents/en/Presentations/2009>.

11 In October 2000, the Chinese government revised *The Law of the People's Republic of China on Foreign Capital Enterprises* to remove the requirement for foreign invested enterprises (FIEs) to export all or most of their products. Later, they also removed the restriction on the domestic sales ratio allowing foreign companies to sell freely to the Chinese market. (In 2009, China's export of goods and commodities across all sectors was US$1.2 trillion. The US is China's largest importer by far, purchasing nearly US$300 billion in 2009.) These are China's Customs statistics as cited by the US-China Business council at <www.uschina.org>.

  Specific Chinese government programs for the timber industry have included low-interest loans; tax exemptions and rebates; and fast-tracking project approvals. American manufacturers have successfully appealed to the International Trade Commission that various forest and paper products from China are being subsidized and dumped on the American market. In response, the US Department of Commerce has placed duties on wooden bedroom furniture, glossy coated paper, and tissue paper exports from China.

12 China National Household Paper Industry Association, at <www. cnhpia.org/en>, as cited in "Household Paper Giants Clean Up Act," *China Daily* (June 8, 2009), at <www.paper.com.cn>.

13 United Nations Economic Commission for Europe (UNECE), *The Importance of China's Forest Products Markets to the UNECE Region* (Geneva: UNECE and the FAO, 2009), 5, 10. Also, for more background on China's wood product industry, see Tropical Forest Trust, *China Wood Products Supply Chain Analysis: Helping Chinese Wood Producers Achieve Market Demands for Legal and Sustainable Timber* (Report prepared by The Tropical Forest Trust

with funding by DEFRA and WSSD Implementation Fund, 2007).

14 Andy White et al., *China and the Global Market for Forest Products: Transforming Trade to Benefit Forests and Livelihoods* (Washington, DC: Forest Trends, March 2006), 3.

15 Examples of Japanese flooring plants in China include: Nitto (Yunnan), Yahima (Dalian), Woodone (Shanghai), Diaken (Ningbo), Josei (Dalian), I.M.S. (Jilin), Wood One Co. Ltd. (Shanghai). Recently, factors such as overcapacity and the appreciating Chinese Yuan have encouraged many wood manufacturers in China to relocate once again to even lower-cost areas, including Vietnam, Thailand, Malaysia, and Indonesia.

CHAPTER 4: THE RISE OF THE THIRD WORLD

1 See "Wood, Pulp & Paper: Trends in Global Production and Trade" (under the tab Industry Sections), available at: <www.global-production.com/wood-pulp-paper/trendstudy/index.htm>.

2 The term "born global" was coined in a 1993 study by McKinsey & Company and the Australian Manufacturing Council. See Michael Rennie, "Global Competitiveness: Born Global," *The McKinsey Quarterly* 4 (1993): 45–52. The definition of born global can range somewhat depending on percentage of foreign sales, innovation, and how and when the firms enter global markets. For details, see S. Tamer Cavusgil and Gary Knight, *Born Global Firms: A New International Enterprise* (New York: Business Expert Press, 2009).

3 Fibria Company Presentation, "Fibria: A Global Leader is Born," São Paulo, Brazil, September 1, 2009.

4 See Jim Bell, Rod McNaughton, and Stephen Young, "'Born-again Global' Firms: An Extension to the 'Born Global' Phenomenon," *Journal of International Management* 7 (3) (2001): 173–89, quote at p. 173.

5 See "Wood, Pulp & Paper," available at <www.global-production. com>.

6 The information in this paragraph draws on: Peter Frandina, David Rossi, and Cyronne Counts, *Trends in Manufacturing Paper Products: Investment Strategies for High Performance in a Multi-polar World* (Acenture Consulting, 2008) (for the projection on

Latin America); Dave Hillman, "China's Pulp Purchases Reach
13.68 Million Mt in 2009," published at <www.paperage.com>
on February 12, 2010, posted on the Gerson Lehrman Group
website at <www.glgroup.com> (for China's imports of pulp and
recycled fiber); and Christopher Barr, *Intensively Managed Forest
Plantations in Indonesia: Overview of Recent Trends and Current
Plans*, presentation at a conference in Pekanbaru, Indonesia,
March 7–8 (Bogor, Indonesia: Center for International Forestry
Research, 2007) (for assessing timber production in Asia).

7 Chris Lang, "Laos: Chinese Company Sun Paper Plans
Eucalyptus Monocultures," *WRM Bulletin* No. 144, July (2009), at
<www.wrm.org/bulletin/144/Laos.html>.

8 PricewaterhouseCoopers (PwC), *Global Forest, Paper & Packaging
Industry Survey: 2009 Edition* (Vancouver, BC: PwC LLP,
2009), 9 (for the 2008 figure); PwC, *Forest, Paper & Packaging
Deals: Branching Out – 2009 Annual Review* (Vancouver, BC:
PricewaterhouseCoopers LLP, 2010), 2, 4.

9 See Fred Cubbage et al., *Global Forest Plantation Investment
Returns*, Presentation to XIII World Forestry Congress, Buenos
Aires, Argentina, October 18–23, 2009; Jacek Siry et al., "Supply
Chain Efficiency: What One Industry is Doing Better," *Forest
Products Journal* 56 (10) (2006): 4–10; Felippe Reis, "Fibria: Still
Dealing with High Leverage," Latin American Equity Research,
Banco Santander S.A., São Paulo, December 7, 2009: <www.
santander.com>.

10 See Jacek Siry et al., "Supply Chain Efficiency," for the estimate
on pulp production from eucalyptus plantations in Brazil; see
Stora Enso, *Annual Report 2007* (Stora Enso, 2007), p. 65, for the
estimate of the growth rate of trees in Scandinavia; see Vincent
Honnold, "Developments in the Sourcing of Raw Materials for
the Production of Paper," *Journal of International Commerce
and Economics* (United States International Trade Commission,
August 2009): 8, for the maturation rates of radiata pine.

11 See the FAOSTAT database at <http://faostat.fao.org>; and
Ronald Gonzalez, Daniel Saloni, Sudipta Dasmohapatra, and
Fred Cubbage, "South America: Industrial Roundwood Supply
Potential," *BioResources* 3 (1) (2008): 255–69.

12 The data in this paragraph are from: Richard English, "Global
Trends in the Forest Products Market," International Finance
Corporation (IFC) Presentation to the India Farm Forestry

Advisory Program Workshop, New Delhi, September 2008 (estimate on role of emerging economies in world paper markets); Pulp & Paper International, *Annual Review*, 2008, and RISI *Global Industry Statistics Database*, 2008, as summarized in Honnold, "Developments in the Sourcing of Raw Materials for the Production of Paper" (US and China's share); Frandina, Rossi, and Counts, *Trends in Manufacturing Paper Products* (role of other developing economies); and English, "Global Trends in the Forest Products Market" (growth rate of Northern economies).

13 RMB, or the Chinese Yuan, is the official currency of the People's Republic of China. For details on the varying profits for Chinese companies producing containerboard, see the website for Exane Derivatives Asia, at <www.exanederivativesasia.com>, May 7, 2007.

14 Peter Dauvergne, *Loggers and Degradation in the Asia-Pacific: Corporations and Environmental Management* (Cambridge, UK: Cambridge University Press, 2001); Michael L. Ross, *Timber Booms and Institutional Breakdown in Southeast Asia* (Cambridge, UK: Cambridge University Press, 2001); Luca Tacconi, *Illegal Logging: Law Enforcement, Livelihoods and the Timber Trade* (London: Earthscan, 2007).

15 Mark Rushton, "APP Indonesia: A Giant Poised on the Edge," *Pulp and Paper International* (December 2009): 16–19.

CHAPTER 5: CONSUMING THE SOUTH

1 John Perlin, *A Forest Journey: The Story of Wood and Civilization*, updated edition (Woodstock, VT: The Countryman Press, 2005), 25.

2 American Forest & Paper Association, "US Forest Products Industry – Competitive Challenges in a Global Marketplace," at <http://www.bipac.net/page.asp?chk=G3Hh xtOky&g=afpa&content=White_Paper_re_Competitiveness_ Summary&parent=AFPA>, June 6, 2005.

3 For the estimate on the percentage of paper in North America and the US going into landfills, see Environmental Protection Agency (EPA), *WasteWise 2008 Annual Report* (Washington, DC: US EPA, 2008), 6; and EPA, *Municipal Solid Waste Generation, Recycling, and Disposal in the United States: Facts*

and Figures for 2008 (Washington, DC: US EPA, 2009), 4. For Walmart's packaging scorecard, see <http://walmartstores. com/Sustainability/9125.aspx>. The EU packaging directive, which requires member states to recover at least 60 percent of its packaging, is available at <http://europa.eu/legislation_ summaries/environment/ waste_management/l21207_en.htm>; the summary of progress is at <www.paperrecovery.org/facts/ erpc_facts_figures.asp?folderid=527>.

4 The "ten-foot high wall" quote in this paragraph is from Annie Leonard, *The Story of Stuff* (New York: Free Press, 2010), 8; the survey is summarized in Louise Story, "Junk Mail is Alive and Growing," *New York Times*, November 2, 2006, available at <nytimes.com>; Jim Ford, *Climate Change Enclosed: Junk Mail's Impact on Global Warming* (San Francisco, CA: Forest Ethics, August 2008). For statistics on junk mail delivery, see United States Postal Service (USPS), *The Household Diary Study: Mail Use and Attitudes in FY 2008* (Washington, DC: USPS, March 2009), 40; for advocacy group statistics on paper consumption and disposal, see <www.forestethics.org> and the Environmental Paper Network at <www.environmentalpaper.org>.

Ljubomir Stambuk conducted the Consumer Research Institute survey in April 2001. For industry paper consumption and recycling statistics, see the American Forest & Paper Association at <http://paperrecycles.org/ stat_pages/stat_ intro_2010.html>. For the statistic on the number of trees needed to produce US junk mail, see Center for an American New Dream calculation at <www.newdream.org/junkmail/facts.php>.

5 Marshall S. White and Peter Hamner, "Pallets Move the World: The Case for Developing System-based Designs for Unit Loads," *Forest Products Journal* 55 (3) (2005): 8–16. For background on the global wood pallet industry, see "The Wooden Crates Organization," at <www.woodencrates.org>; also see Peter Hamner, *Pallets: Where Form Meets Function*, Material Handling Industry of America, October 2007, at <www.mhia.org/news/ industry/7053/pallets-where-form-meets-function>.

Wooden cargo crates also contribute to the waste stream, but the volume is smaller than for pallets. A crate is a custom-designed six-sided container; a pallet is flat and has international standardized measurements. See Jeff Duck, "Pallets v. Crates: Searching for Synergies in Two Different Industries," *Pallet*

Enterprise, 2010, at <www.palletenterprise.com/articledatabase/view.asp?articleID=3019>.

6   Urs Buehlmann, Matthew Bumgardner, and Tom Fluharty, "Ban on Landfilling of Wooden Pallets in North Carolina: An Assessment of Recycling and Industry Capacity," *Journal of Cleaner Production* 17 (2) (2009): 271–2; Robert Bush et al., "Trends in the Use of Materials for Pallets and Other Factors Affecting the Demand for Hardwood Products," Proceedings of the 30th Annual Hardwood Symposium, Tennessee, May 30–June 1, 2002; "Hardwood Market Report, 2006: The Year at a Glance," in *10th Annual Statistical Analysis of the North American Hardwood Marketplace*, Memphis, 2007, p. 172; Bob Smith and Victor Cossio, *Competitiveness of Forest Products at Global Markets: Market Review in the US of Selected Timber Products*, Department of Wood Science & Forest Products, Virginia Tech, January 23, 2008, for the Food and Agriculture Organization, available at <www.fao.org/forestry/18280-1-0.pdf>; J.D. Piland, "Wood Pallets Wasted no More," *Wood & Wood Products* (July 2004): 171–8.

7   Ingvar Kamprad, "A Furniture Dealer's Testament," first published in 1976, in Bertil Torekull, ed., *Leading by Design: The IKEA Story* (New York: HarperCollins, 1994) as cited on the IKEA website at <www.ikea.com/ms/en_GB/about_ikea/press_room/student_info.html>.

8   See William F. Laurance et al., "The Need to Cut China's Illegal Timber Imports," *Science Magazine* 29 (February) (2008): 1184–5. As well, for analysis and estimates of China's import of illegal timber, see the website for Global Timber at <www.globaltimber.org.uk/ChinaIllegalImpExp.htm>.

9   ITTO, *Annual Review and Assessment of the World Timber Situation* (Yokohama: International Tropical Timber Organization, 2008), 9. The ITTO Annual Reviews are a reasonable (though somewhat cautious) source for statistics on tropical timber and are available at <www.itto.int/en/annual_review>.

10 WWF, *Illegal Wood for the European Market: An Analysis of the EU Import and Export of Illegal Wood and Related Products* (Frankfurt, Germany: World Wildlife Fund, 2008), 14.

11 See Kavita Watsa, *Commodities at the Crossroads: Key Findings from Global Economic Prospects 2009* (Washington, DC: International Bank for Reconstruction and Development and World Bank, 2009), 2.

12 This paragraph draws on Mitsui & Company Europe, "Multigrain: Meeting Surging Global Demand for Grain," at <www.mitsui.eu/business/challenge/grain/index.html>; Jan Willem van Gelder and Jan Maarten Dros, *From Rainforest to Chicken Breast: Effects of Soybean Cultivation for Animal Feed on People and Nature in the Amazon Region – A Chain of Custody Study*. A research report for the Dutch Soy Coalition commissioned by Friends of the Earth and Cordaid, Netherlands, January 17, 2006; Maria del Carmen Vera-Diaz, Robert Kaufmann, and Daniel Nepstad, "The Environmental Impacts of Soybean Expansion and Infrastructure Development in Brazil's Amazon Basin," Global Development and Environment Institute, Tufts University, Medford, MA, May 2009; Nina Holland et al., *The Roundtable on Ir-responsible Soy: Certifying Soy Expansion, GM Soy and Agrofuels*, April 2008, at <http://archive.corporateeurope.org/ docs/soygreenwash.pdf>. For statistics on the global trade in soybeans, see US Foreign Agriculture Service, *Oilseeds: World Markets & Trade*, May 2010, at <www.fas.usda.gov/oilseeds/circular/Current.asp>.

13 The phrase "deforestation diesel" is in Niluksi Koswanage, "Palm Oil CO_2 Targets Delayed as Planters, NGOs Clash," *Reuters*, November 2, 2009.

The Food and Agriculture Organization projects that biomass energy use in Europe will triple by 2020 in response to the EU's 20 percent renewable energy target. See FAO, *State of the World's Forests* (Rome: FAO, 2009), 68. For analysis of the emerging political economy of biofuels, see Peter Dauvergne and Kate J. Neville, "The Changing North–South and South–South Political Economy of Biofuels," *Third World Quarterly* 30 (6) (2009): 1087–1102; Peter Dauvergne and Kate J. Neville, "Forests, Food, and Fuel in the Tropics: The Uneven Social and Ecological Consequences of the Emerging Political Economy of Biofuels," *The Journal of Peasant Studies* 37 (3) (2010): 631–60.

14 For statistics on deforestation, see FAO, *State of the World's Forests 2009*; M. Hansen et al., "Humid Tropical Forest Clearing from 2000 to 2005 Quantified by Using Multi-temporal and Multi-resolution Remotely Sensed Data," *Proceedings of the National Academy of Sciences* 105 (27): 9439–44.

For analysis of the impacts of logging on community security, see Richard A. Matthew and Ted Gaulin, "Conflict or Cooperation? The Social and Political Impacts of Resource

Scarcity on Small Island States, *Global Environmental Politics* 1 (2) (2001): 48–70; Peter Dauvergne, "Corporate Power in the Forests of the Solomon Islands," *Pacific Affairs* 71 (4) (winter 1998–9): 524–46.

15 See Yumiko Uryu et al., *Deforestation, Forest Degradation, Biodiversity Loss and CO2 Emissions in Riau, Sumatra, Indonesia* (Jakarta: WWF Indonesia Technical Report, February 27, 2008).

CHAPTER 6: GOVERNING TIMBER CONSUMPTION

1 Christine MacDonald, *Green Inc.* (Guilford, CT: The Lyons Press, 2008), 151.

2 See the website for the Sustainability Consortium, at <http://sustainabilityconsortium.org>.

3 Quoted in Colin Sullivan, "Wal-Mart's Chairman Pulls a Long Supply Chain toward Sustainability," *ClimateWire* (Washington, DC: Environment & Energy Publishing LLC, 14 April, 2010), at <www.eenews.net/public/ climatewire/print/2010/04/14/1>.

4 Chris Lazlo, *Sustainable Value: How the World's Leading Companies are Doing Well by Doing Good* (Sheffield, UK: Greenleaf Publishing, 2008).

5 See the Home Depot website, Wood Purchasing Policy, at <http://corporate.homedepot.com/wps/portal/Wood_Purchasing>.

6 Examples of illegally traded endangered tree species include: African blackwood, cherry and mahogany; rosewood from Madagascar; teak from Burma; Honduras mahogany; and kawi and keruing from Indonesia. See the United Nations Environment Programme World Conservation Monitoring Centre (at <www.unep-wcmc.org/forest/timber/Default.aspx>) for a complete record of the Convention on International Trade in Endangered Species (CITES) Annex I, II, and III listed tree species.

7 See the Forest Footprint Disclosure Project website, at <www.forestdisclosure.com>.

8 See the Carbon Disclosure Project website, at <www.cdproject.net> (under "more about CDP").

9 IKEA, *IKEA Sustainability Report* (IKEA, 2008), 33.

10 See the Staples website, Environmentally Preferably

Products, at <www.staples.com/sbd/content/about/soul/environmentallypreferableproducts.html>.

11 IKEA conducted 60 wood supply chain audits in 2009. See IKEA, *IKEA 2009 Sustainability Report*, p. 58, for a listing of the company's forestry requirements.

12 Statement as cited in Staples, *Staples Soul Report* (Staples, 2007), 11.

13 See Peter Dauvergne and Jane Lister, "The Prospects and Limits of Eco-Consumerism: Shopping Our Way to Less Deforestation?," *Organization & Environment* 23 (2) (2010): 132–54.

14 Quoted in Colin Sullivan, "Wal-Mart's Chairman."

Selected Readings

Readers wanting to broaden the analysis in chapter 1 and learn more about the history of forests as a world resource may want to start with John Perlin, *A Forest Journey: The Story of Wood and Civilization* (Woodstock, VT: The Countryman Press, 2005, updated edition), and Michael Williams, *Deforesting the Earth: From Prehistory to Global Crisis*, An Abridgment (Chicago, IL: University of Chicago Press, 2006). For an excellent overview of the governing crisis surrounding the political economy of deforestation, see David Humphreys, *Logjam: Deforestation and the Crisis of Global Governance* (London: Earthscan, 2005). For overarching analysis of the role of big business in shaping the political economy of the global environment, see Doris Fuchs, *Business Power in Global Governance* (Boulder, CO: Lynne Rienner, 2007), and Jennifer Clapp and Doris Fuchs, eds., *Corporate Power in Global Agrifood Governance* (Cambridge, MA: MIT Press, 2009). The United Nation's Food and Agriculture Organization (<www.fao.org>) is a good starting point for statistics (e.g., see the FAO Stat database for global forest area, production, trade, and consumption at <http://faostat.fao.org/site/630/default.aspx>), but reporting difficulties, definitional differences, and institutional biases mean it is best to compare estimates across other sources, such as the World Resources Institute (<www.wri.org>), the International Tropical Timber Organization (<www.itto. org>), and the many governmental, non-governmental, and corporate organizations tracking and documenting timber.

Chapter 2's analysis of the rise over the last two decades of big retail and the world discount economy has been a source of many recent popular and academic books. As a sampling, see Gordon Laird, *The Price of a Bargain* (Toronto: McClelland & Stewart, 2009), Michael Andersen and Flemming Poulfelt, *Discount Business Strategy: How the New Market Leaders are Redefining Business Strategy* (Chichester, UK: John Wiley & Sons, 2006), Stacy Mitchell, *Big-Box Swindle* (Boston, MA: Beacon Press, 2006), and Ellen Ruppel Shell, *Cheap: The High Cost of Discount Culture* (New York: Penguin, 2009). For an academic-oriented book on Walmart, see Stanley D. Brunn, ed., *Wal-Mart World: The World's Biggest Corporation in the Global Economy* (London: Routledge, 2006). For a popular book on Wal-Mart's role in this discount economy, see Charles Fishman, *The Wal-Mart Effect: How the World's Most Powerful Company Really Works – and How It's Transforming the American Economy* (New York: Penguin, 2006). Academic analysis of the growing importance of global commodity chains is also developing quickly. A foundational book is Gary Gereffi and Miguel Korzeniewicz, eds., *Commodity Chains and Global Capitalism* (London: Praeger, 1994); for a sampling of more recent analysis, see Jennifer Bair, ed., *Frontiers of Commodity Chain Research* (San Francisco, CA: Stanford University Press, 2009). For statistics on the global retail industry and economy, see among many other sources, the annual reports prepared by the large global accounting firms: Deloite Touch Tohmatsu, *Global Powers of Retailing*; KPMG International, *Global M&A Outlook for Retail*; and A. T. Kearney, *Global Retail Development Index* regarding emerging retail market strategies and opportunities.

More general coverage of chapter 3's analysis of the shifting role of multinational corporations in global environmental governance can be found in David Levy and Peter Newell, eds., *The Business of Global Environmental Governance* (Cambridge,

MA: MIT Press, 2005). For a more specific discussion of the importance of conflict and competition among MNCs, see Robert Falkner, *Business Power and Conflict in International Environmental Politics* (London: Palgrave Macmillan, 2007). There is a large business and economics literature on forest resource management, domestic forest economies, the structure of the global forest industry, and the competitiveness of multinational timber companies. For forest resource economics, see, for example, G. Cornelis van Kooten and Henk Folmer, *Land and Forest Economics* (Cheltenham, UK: Edward Elgar, 2004). For an analysis and modeling of international trade in forest products, see Joseph Buongiorno, Shushuai Zhu, Dali Zhang, James Turner, and David Tomberlin, *The Global Forest Products Model: Structure, Estimation and Applications* (San Diego: Academic Press, 2003). And for a comparative overview of changing firm-level strategies in the global forestry sector, see J. A. Lamberg, J. Näsi, J. Ojala, and P. Sajasalo, eds., *The Evolution of Competitive Strategies in Global Forestry Industries: Comparative Perspectives* (Dordrecht, The Netherlands: Springer, 2006). Far less has been written on the political economy of Northern multinational timber firms. A useful starting point is Ari Aukusti Lehtinen, Jakob Donner-Amnell, and Bjørnar Sæther, *Politics of Forests: Northern Forest-industrial Regimes in the Age of Globalization* (Aldershot, UK: Ashgate, 2004).

Background on chapter 4's analysis of the rise of Third World firms within the global economy, especially the increasing importance of the trend among these firms to be "born global," can be found in S. Tamer Cavusgil and Gary Knight, *Born Global Firms: A New International Enterprise* (Williston, VT: Business Expert Press, 2009). Scholarship on the specific practices of Third World timber firms is extensive. For a sampling, see Peter Dauvergne, *Loggers and Degradation in the Asia-Pacific: Corporations and Environmental Management*

(Cambridge, UK: Cambridge University Press, 2001); Michael L. Ross, *Timber Booms and Institutional Breakdown in Southeast Asia* (Cambridge, UK: Cambridge University Press, 2001); and Luca Tacconi, *Illegal Logging: Law Enforcement, Livelihoods and the Timber Trade* (London: Earthscan, 2007). For an analysis of international efforts to manage trade in tropical timber, see Fred P. Gale, *The Tropical Timber Trade Regime* (London: Macmillan, 1998).

For a big picture overview of chapter 5's analysis of the shadow effects of consumption on the global South, see Peter Dauvergne, *The Shadows of Consumption: Consequences for the Global Environment* (Cambridge, MA: MIT Press, 2008). For a more detailed study of the effect of Japan on timber management in Southeast Asia, see Peter Dauvergne, *Shadows in the Forest: Japan and the Politics of Timber in Southeast Asia* (Cambridge, MA: MIT Press, 1997). For an accessible overview of the historical and modern role of global food networks in driving environmental change worldwide, see Evan D. G. Fraser and Andrew Rimas, *Empires of Food: Feast, Famine, and the Rise and Fall of Civilizations* (New York: Free Press, 2010).

Chapter 6 is only able to briefly touch on the complex relationship between corporate social responsibility and global forest governance. For an overview of the definition, challenges, and opportunities of corporate social responsibility, see Bryan Horrigan, *CSR in the 21st Century: Debates, Models and Practices across Government, Law and Business* (Cheltenham, UK: Edward Elgar, 2010); David Vogel, *The Market for Virtue: The Potential and Limits of Corporate Social Responsibility* (Washington, DC: The Brookings Institute, 2005); Jane Nelson, *The Public Role of Private Enterprise: Risks, Opportunities and New Modes of Engagement* (Boston, MA: Harvard University Press, 2006); and Simon Zadek, *The Civil Corporation: The New Economy of Corporate Citizenship* (London: Earthscan, 2001). For an understanding of emerging green marketing and global eco-business strategies, see Chris Arnold, *Ethical Marketing and the New Consumer: Marketing in the New Ethical Economy* (Chichester, UK: John Wiley & Sons, 2008); Kellie McElhaney, *Just Good Business: The Strategic Guide to Aligning Corporate Responsibility and Brand* (San Francisco, CA: Berrett-Koehler Publishers, 2008); John Grant, *The Green Marketing Manifesto* (Chichester, UK: John Wiley & Sons, 2007); Neil Stern and Willard Ander, *Greentailing and Other Revolutions in Retail: Hot Ideas That Are Grabbing Customers' Attention and Raising Profits* (Chichester, UK: John Wiley & Sons, 2008); and Tim Divinney, Pat Auger, and Giana Eckhardt, *The Myth of the Ethical Consumer* (Cambridge: Cambridge University Press, 2010). Particularly valuable books for those wanting to understand further the relationship of CSR to forest governance include: Lars H. Gulbrandsen, *Transnational Environmental Governance: The Emergence and Effects of the Certification of Forests and Fisheries* (Cheltenham, UK: Edward Elgar Publishing, 2010); Benjamin Cashore, Graeme Auld, and Deanna Newsom, *Governing through Markets: Forest Certification and the Emergence of Non-state Authority* (New Haven, CT: Yale University Press, 2004); and Jane Lister, *Corporate Social Responsibility and the State: International Approaches to Forest Co-Regulation* (Vancouver, BC: UBC Press, 2011). A useful resource comparing environmental forest policies and enforcement across 20 countries is Constance McDermott, Benjamin Cashore, and Peter Kanowski, *Global Environmental Forest Policies: An International Comparison* (London: Earthscan, 2010). For those wanting a broader discussion to explore "decarbonizing" the global economy, and how this might relate to corporate social responsibility and global environmental governance, see Peter Newell and Matthew Paterson, *Climate Capitalism: Global Warming and the Transformation of the Global Economy* (Cambridge, UK: Cambridge University Press, 2010); also see Ricardo Bayon, Amanda Hawn, and Katherine Hamilton, *Voluntary Carbon Markets: An International Business Guide to What They Are and How They Work* (London: Earthscan, 2009); and Anja Kollmuss et al., *Handbook of Carbon Offset Programs* (Washington, DC: Earthscan, 2009).

Index